THE DARWIN CONSPIRACY

Revised Edition

I0462808

**The Conspiracy that Gave Darwin
the Title of The Father of Natural Selection**

by

**YUVENALIY VLADIMIROVICH
CLADOVAYNIKOFF**

(English Version – Ivan Kalinskiy)

Moscow, 2009

ISBN-10:   1448699819

ISBN-13:   978-1448699810

CONTENTS

Preface to Revised English Version
by Ivan Y. Kalinskiy

Since this book was originally published, many new facts concerning Darwin's shadowy character and facts about the conspiracy have come to light. A revised edition was essential. While most writers choose to continue to join the other "keepers-of-the-flame" of Darwin's fame and the Father of Natural Selection despite the most recent findings that not only *confirm* the circumstantial evidence we originally published, but outline how he perpetrated the conspiracy. It allowed him to be called "the Father of Natural Selection" and later "Father of Evolution." This book is more essential than ever to tell the whole true story.

I would like to mention, that I am a neurophysiologist, not a writer, nor well versed in punctuation, grammar, syntax, structure or style. Therefore the original English version was filled with such errors for which I am responsible and I apologize. However, even with my best efforts, the same may still be true because the publisher did not edit the text. We felt that a revised edition to correct these flaws was in order. Also some paragraphs were disjointed and needed to be repositioned and others expanded or clarified.

Secondly, there was a time constraint. I had to complete the first publishing before I returned to Russia. The first publisher created so many delays in working with the two books I entrusted to them in 2007 that I despaired of seeing anything started by 2009 and had to find another publisher at the last minute. The new publisher saved the day, but could do it only by printing what was there without editing; consequently, there was no editing and my errors were printed as it appeared. However, I have the highest praise for publisher for having completed everything so rapidly. It was a revelation that a publishing house could work so fast and I was able to take copies back to Russia with me to show Yuvenaliy Valdimirovich Kladovainikov. While the book spelled his name with a "C", we have not changed this error to have a consistent listing and reference. Many Russian names are misspelled when translated from the Cyrillic alphabet into the Latin.

Thirdly, the Russian edition should have come out first but the delay here can be even more extreme than overseas where in America the entire book is handled electronically. In Russia at present, the opposite is true. The number of copied, the distribution, and editing have to be decided first and the cost can be extremely high before one sees the first copy. If any errors in printing occur, the entire edition is in your hands!

Fourthly, Yuvenaliy Vladimirovich (Kladovainikov) wanted to make changes to clarify issues that were raised after the first edition was made public, and to make the text more readable by rearranging several paragraph and adding several more.

Most important, there are new facts recently made public which confirms that a conspiracy took place. The book contained conjectures based on previous circumstantial evidence *but the facts now are established that a conspiracy was perpetrated and are included.*

Surprisingly, this period in history when the conspiracy occurred is receiving far more attention today than at any time previously. It is probably due to more unbiased research, recent articles and hopefully this book, which is resulting in the down-sizing of Darwin and the increased prestige of Alfred Russel Wallace. In addition Yuvenaliy Vladimirovich[1] has uncovered a condition that Darwin had, which was strangely unnoticed by previous medical analyses or writers earlier, despite the number of previous diagnoses and analyses of his life and behavior. They merely concentrated on his complaints than on his behavior. The complaints were more of a sign of the problem than the cause.

Ivan Kalinskiy

---

[1] Note: It is common to use the first name and the father's first name and not the last name when addressing or speaking about a person one knows.

INTRODUCTION

This text grew over the years from my first reading of the Harvard Classics[2] which includes the *On the Origin of Species* by Charles Darwin (1809-1882), particularly the Introductory Notes which contains some very strange facts about the first reading of the paper on evolution before the Linnaean Society **on July 1, 1858**. Remember, **before this date, at age forty-eight, Darwin had not published one word on evolution.** We have only his word, notes and letters about what he **may have done.** We will refer to another pro-Darwinist who accepted that he was a "braggadocios" type and from the facts of his life, we know that he was not above "self-gratuitous fibs" about what he may or may not have done.

On **June 16th of that year (only two weeks earlier)** Darwin had received a full manuscript on natural selection from Alfred Russel Wallace requesting him to present his paper on evolution to the prestigious Linnaean Society since Darwin was a very wealthy and influential member and it might carry more weight. Darwin had no manuscript to present because he had

---

[2] This text was written earlier prior to 2007 but published only in 2009 and now updated in 2010. After the first edition was published many new facts about his less than honorable character and detailed information on this conspiracy were uncovered.

1

stopped working on evolution in 1850 but rather was preoccupied with an extensive classification of barnacles and other papers not associated with evolution (e.g., manuscripts on geology, effects of salt water on seeds, etc.) *for the previous eight years*! [Note: Had he found the subject of evolution more involved than he had previously thought or *for whatever reason*, he switched to classifying barnacles. Or were his "long reflections" on evolution going nowhere?].

[Note: Darwin had taken only one trip around the world, mostly at sea, writing about his adventure, and some geological observations (see below). *Before, during and for some years afterwards, he was preparing to be a clergyman* in the Church of England and had no interest in science let alone evolution because he strongly believed in Biblical creation. At age thirty-three he secluded himself with his wife, family and servants in his mansion in the little village of Downe, Trent to tend to gardening and writing letters and doing "research." Such was the full extent of Darwin's experience with evolution.

[I*t seems* that Darwin set up this conspiracy, using Wallace's paper, without the other co-conspirators knowing the extent of it. The other two conspirators apparently believed that in fact Darwin had his own paper *which just happened to be almost identical to that of Alfred Russel Wallace's paper!* The three conspirators

2

influenced the president of the Linnaean Society to become involved in the conspiracy. It did not appear that Darwin had any intention at that time to present a paper, nor even had a paper to present and was certainly not ready to publish anything on evolution which he stopped working on eight years earlier.]

To me, this smacked of some underhanded maneuvers to ensure that Darwin got all the credit, and he did. I continued to read the limited sources that were available to me which further roused my suspicions about how this presentation was handled where Darwin, coming from behind, ended up first!

Most sources are so emphatically in favor of eulogizing Darwin it was hard to get an accurate picture of the man or what actually happened. Different sources did present some facts that further confirmed to me that something dishonest *had* been perpetrated, despite these same sources showering praises on Darwin. From recent articles, more objectively written, but still somewhat guarded in saying anything really defamatory about Darwin, they did more candidly speak about the facts that related to this incident.

This present book is my compilation of facts leading up to the events surrounding the presentation at the Linnaean Society where the theory of natural selection was first read.

Many of the fact surrounding the presentation at the Linnaean Society have been known for years, but have been ignored for whatever reason. It seemed like the greater Darwin's reputation, the greater the desire to white-wash his unwarranted conjectures and in some cases to try and cover up or excuse his major mistaken ideas, pass over numerous improprieties, (and as I feel, downright fabrications), as well as this conspiracy. So many superlatives have been added to his name and eulogies for the man as well as his work that we would undoubtedly not recognize the real person *or at least, not the man described below which comes directly from previous references.*

Many Russian revolutions started with a bomb-thrower and many believed that we carried bombs daily a la Trotskyites (Comrade, I think my bomb is slow, what time is it by *your* bomb?).[3] This book may be considered bomb throwing at the most revered idol in biological science. To explode the reputation of such a demigod of science is pure heresy, (but if is true...?). Now we have the facts!

---

[3] However, not all Russian revolutions were violent. Maleevich, Kandinsky and Shevchenko started the revolution in art by creating the first abstract art. Tatlin at that time started a revolution in architecture (constructivism) and in 1910 Igor Stravinsky created a revolution in music with one composition, the radical "*Rite of String*" which changed modern classical music thereafter.

Many idols of the past have been bombed or attacked (Columbus' reputation had been exploded for bringing small pox to the Americas. Stalin was blasted by Khrushchev for his reign of fear and even Jesus of Nazareth has had his "bomb-throwing" attackers as well). Objectively, anyone can come under attack, and Darwin himself could *hardly* cast the first bomb. Both Darwin and Stalin, started out to be clergymen and later lobbed bombs at their rivals, particularly in the direction of the church and help to destroy the concept of a deity more than any other two men. Stalin's bombs were aimed to kill people and literally destroy churches, but Darwin's bombs aimed at Lamarck, Chambers and Milne and others were just as lethal. But what Darwin did underhandedly was even worse; it changed history in unimaginable ways [though perhaps it would have been inevitable anyhow since he was only a sign of the times, not the mover of events].

In his defense, Darwin *did* **inadvertently** promote evolution through his, wealth, unique position in high society and influential friends in the scientific community. Other authors lacked these contacts and never received the fame or credit that they deserved for their hard work on evolution.

However, to have denied the rightful claim of Alfred Russel Wallace to the title of Father of Natural Selection is going too far. Alfred Russel Wallace (or possibly other who wrote theories on natural selection before Wallace, had every right to this title and I find the facts have been known for over a century. However the pro-Darwinists have managed to hide them and skip over, *or leave out* the essential details which would have made the conspiracy more obvious. It is my interpretation of the facts but proof has already been published elsewhere [finally!] and my desire to give Alfred Russel Wallace the respect he deserves.[4]

I.   References

This book is not a research paper where every word I state is documented. However, just about every quote and factual material can be found in the reference I used. *Where I interject comments, I try to put them in brackets "[ ]" without further comments as the one immediately below.*

[Note: There is probably no man in history where there is such a variety of accounts about any part of his life or works. Much of it is due to Darwin himself who not only reported the facts differently at different times

---

[4] My biochemistry professor was rumored to be a great "might-have-been" had he had the time to present his work on the pancreas, but was held up due to the war years. At the same, time, Banting, Best and Macleod in Canada published their work on insulin and became world famous. Had it not been for this brief intervention caused by pressing demands due to the war, he would have published first, but didn't. His name faded from history. In the present case, Alfred Wallace **was first** but Darwin, Sir Charles Lyell and the scientific elite of the day took it away from him.

but also when he *thought of something* or did things. In all instances it appears that such changes were in his favor or changes in his theory were opportune. For example, his book about the *Voyage of the Beagle (1839)* started out as primarily an adventure story and geological observations as a result of a book he had read on the trip on geology by Sir Charles Lyell. The next edition 1845 strangely changed to one of natural history and hints of evolution as the theories of evolution raged about him and he had a chance to obtain more information about evolution and natural history from his friends. Geology was eclipsed. He gives three different dates for his interest in evolution and a fourth is more likely. Yet, researchers seem to slide over these changes as if they were invisible. They take Darwin's word as fact as if they were Talmudic scholars when reading the Old Testament.[5]]

II.   Commenting on my Sources

Lacking access to Darwin's letters, and memoirs (and wondering, *for good reason*, how much is fact or "gratuitous fibs"),[6] I have relied heavily on several other sources for details surrounding this conspiracy. I base my

---

[5] These scholars when faced with total inconsistencies when reading the Talmud have to see them as indisputable facts that needs to be accepted, never questioned, such as the three different sets of Ten Commands in the Bible. They can interpret but not change a word.

[6] Encyclopedia Britannica Vol. 5, p492-3, 1959 Ed.)

conclusions on what is written in Darwin's *On the ORIGIN of SPECIES BY NATURAL SELECTION or the PRESERVATION OF FAVORED RACES IN THE STRUGGLE FOR LIFE,* itself (1909-14 edition of the Harvard Classics), **particular the Introductory and Historical notes** that precede the actual book, which was my first inkling of this conspiracy. I have been troubled about it ever since, but had neither the time nor resources to further investigate the issues. The *Encyclopedia Britannica* (15th Edition, 1959 Vol. 5 p-492-3) gives further details but still adds profuse editorial compliments to Darwin while denigrating Wallace with a brief description of the man and his work and in one case interjecting a snide remark that seemed to shed doubt on Wallace's veracity while defending Darwin (Vol. 19 p-530).[7] [Note" The Encyclopedia Britannica has changed their opinion of Alfred Russel Wallace radically in the latest deluxe edition dated 2010 (available in 2009). An excellent recent article in the *Smithsonian Magazine,* June 2008, by Richard Conniff entitled: *On the Origin of a Theory,* gave additional insight into the situation. It is well worth reading. A few articles in the *National Geographic Magazine* on Wallace (Dec 2008) and another by David Guamman: Feb 2009 on Darwin gives

---

[7] Recently, in the latest version of the Encyclopedia Britannica Deluxe edition Chicago 2010 (?) (It was still 2009!). They have completely reversed themselves and given Alfred Russel Wallace some belated credit and essentially admitted that an underhanded trick was played on Wallace.

some additional facts which mentioned the confusion (or *misrepresentations?*) on the part of Darwin in reporting about his first interest in evolution. Then there was the late Patriarch of Darwinism, Harvard Professor Steven Jay Gould who wrote *The Structure of Evolutionary Theory* a four hundred page text trying in every way to justify everything that Darwin did including eulogizing his prose that deeply inspired him. He even went so far as to suggest that Darwin should be called the Father of Genetics because he used the word "Pangenesis" (from "genesis" in the Bible, meaning "origin") in a sentence unrelated to genes which were unknown at that time. Darwin, in one of his wild speculations felt that traits were passed through the blood by "gremmules" and landed in the testicles; (a speculation that most Darwinists try to hide). However, Professor Gould does not attribute the title of Father of Eugenics to Darwin which very much applies. The theory of natural selection is the basis of eugenics and continues to be the basis for work on eugenics today. Until I read this book, I felt that Professor Gould was the finest, clearest, most *objective* and *thorough* author that I had ever read in science and had the greatest respect for him. Therefore I was totally taken back by the fact that he became so mesmerized with Darwin that he seemed to have lost all objectivity in his excuses and praise of Darwin. Other recent articles had continued to heap more and more praise on Darwin as if he were some sort of demigod. [In fact, Darwin needs only be canonized to complete the sequence toward his complete divinity.]

Perhaps the most endearing **and inaccurate** article, though there are many, on Darwin is the Introductory and Historic notes in the Harvard Classics which throw compliments around like confetti, such as: "…with Henslow, the professor of botany, led (to) his (Darwin's) enlarging his general scientific knowledge…." (Yet Darwin admits he learned nothing and had no scientific knowledge to enlarge and states that on becoming "chief naturalists" (which he wasn't) on the HMS Beagle, when he admitted that he didn't even know what the word "science" meant!). [But wait, the eulogies (in the Notes) and non factual material become even more ecstatic and inaccurate later!]

III.   Comments about my Book

Despite the above sources, I want to emphasize that I am responsible for calling the famous incident *a conspiracy* although other have reported the incident before. I have put together several telling facts and pose several unanswered questions on which to base my opinion. Others have stated the facts too, almost as clearly, but never given it a name, but left any conclusions to the reader. One can see how this has been done by the authors: briefly stating the facts, then rapidly rushing to emphasize the enormous work Darwin had done *previously. [Hearsay?]* Darwin claims that his entire writings on evolution were always done *"in secret"* in preparing a manuscript. In short, we are expected to accept what he says *by accepting Darwin's word on faith as there was no other proof until after 1858 when he already had the answers to evolution from Wallace, despite Darwin's history of bragging to enhance his self esteem and telling gratuitous" fibs" [a kind way of putting it].* Darwin frequently referred to his "yet unpublished manuscript" as fact. [Me thinks Darwin protesteth too much.][8] However, when he reflected on this "yet unpublished document," the dates strangely changed in his favor. Later, documents were found of preliminary sketches on evolution [but to my knowledge, it is unclear what they uncover about a full theory of evolution.] Darwinists state that this may have been left for posterity in case he never finished his major work, but in that case if someone else had published the theory before him, who would care what he wrote in secret?

I should mention that in Russia very few are interested in evolution for several reasons: the Russian people are very religious; Stalin foisted his pet evolutionist Lysenko on the country as the real theory of evolution and the economy keeps everyone busy with other matters, such as survival. Therefore, the English version is more important but even in America, the

---

[8] Elizabethan English at the time of Shakespeare.

percentage of people interested in evolution is only 39 percent. It may be a tempest in a teapot but one that should be of interest to those that follow evolution and look for the truth.

Y. Kladovainikov, 2010, Moscow

PART ONE:
BACKGROUND TO THE DARWIN CONSPIRACY

I.   Background: Darwin's Wealth and Position in Society

In view of the past eulogies about Darwin and his genius that hide his many unfavorable character traits and odd ideas and why there was a conspiracy in the first place, one has to know the facts about him that show the likelihood of Darwin perpetuating such an act. Otherwise it sounds like a shot in the dark that is incomprehensible. Despite the fact that there is proof that it occurred, many will still deny it happened or blame others. For this reason, a lengthy background on Darwin, his period and thinking on evolution need be presented to give a full picture of the state of evolution when this conspiracy occurred which has been artfully hidden despite the facts.

Less commented on, (**and *essential* to this conspiracy and to the fame of the theory**), is the fact that Darwin was born into a very rich, famous and influential family. He had the best of everything; servants, schooling and never had to earn a kopeck for working a day in his life. He was a man of luxury and leisure with all the time in the world to do what he wanted. His father was already a well known physician and his grandfather, even more famous, for his book "The Botanic Garden." Darwin's father had married a wealthy daughter from the famous Wedgwood (porcelain) family. Therefore money and excellent contacts in the field of science, religion and other disciplines were at his disposal. *In short,* the young *Darwin* **was part of the cream of British society of his time and he graciously acted the part, [in public].** His family name and wealth not only open doors in society for him but to prestigious schools and institutions such as The Royal Society and Linnean Society. **His elite status in high society and his wealth are essential for understanding subsequent developments surrounding the fame of his books and the conspiracy, which permitted it to succeed.** But let us look at *the man himself.* Perhaps *this* portrait of the man is a slanted, but factual view. However, after so many superlatives written about Darwin that overshadow the facts, it is time for a more sobering picture.

II.   Charles Darwin (1809-1882)

Darwin was born in Shrewsbury, England. His mother died when he was eight and was raised by his father, two sisters, a nanny, servants and all the comforts that wealth could provide. He lived in a very class-conscious society in England of which he was part of the elite.

Darwin was described as "developing slowly and given *as a boy to* "***inventing gratuitous fibs and to daydreaming***." (The latter emphasis is my own).[9] [What a polite way to say he was a slow learner, indolent, disinterested in reality and living in fantasy, *given to self congratulatory untruths to build up his ego!*] He collected stamps and pebbles which historians have added: "passionately." The same edition has justified his collecting as "an important trait for a future naturalist."[10] [However, in other descriptions of Darwin little mention is made of his habit of *self congratulatory remarks, twisting the truth and **inventing** gratuitous fibs* which would continue to benefit him *even more* as time went on.]

One would never pick up his personal habits from the image presently described in many biographies and journals because ***after having found fame, it was appropriate for men of outstanding wealth to act humble and gracious about the fact. However, he continued to falsify facts even in his final memoirs.***

It was the custom of the day for men of outstanding wealth to have flattering and dignified portraits painted of themselves. Darwin was no exception. We see him as a statuesque man of distinction, as is often reproduced from his portrait at the British National Portrait Gallery

---

[9] *Encyclopedia Brit.* Vol. 5 p-492 1959 Ed.
[10] *Encyclopedia Brit.* Vol. 5 p-492 1959 Ed.

which is the way we usually see Darwin. [*Until today?*] However, on the other hands, *Wallace has only a tinted photograph to represent him.*

Actually, Darwin was very short, had Neanderthal-like brow ridges and was not very attractive. [Much was made of his appearance in cartoons comparing him to an ape.] He seems to have compensated for his unattractive appearance by developing a superiority complex to build up his ego. [Napoleon complex?] This may have led to his negative traits of bragging about things of questionable fact which his friends took for granted. As we will see below, he was not above revising documents to support his claims. Despite his apparent gentile nature, he was not above being cocky and arrogant, hurling bombs at those he resented or who criticized him.

Darwin is often represented in portraits and photographs as having an unkempt beard because it is claimed it hurt him to shave as he aged, despite the fact that he had the best barbers (perhaps servants) to do it for him. However, as he aged the heavy eyebrows, covering his protruding brow and his long beard made him look like a sage and he may have enjoyed this look and had his portrait painted that way. He died at age 73 which is not all that old. [Many friends of mine shave and are at least ten years older than Darwin was when he died, yet they do not scream in agony when a barber scrapes their skin with his razor. Perhaps Darwin was overly sensitive having been pampered all of his life, or he liked hiding behind his beard and his eyebrows. At least he could have had his beard, eyebrows and hair *trimmed*.]

Younger portraits show him clean shaven and full face toward the painter so that his protruding brow ridges were unnoticeable.

III. Darwin's Education

At the age of 9, he was shipped off to the elite Shrewsbury School only to complain that he was only taught the *classics* where he *must have read Aristotle's theory on* **natural selection** (the *first* such theory ever written) but seems to have shown absolutely no interest. He was such a poor student that even at sixteen his father scolded him saying: "You care for nothing but shooting, dogs and rat catching and you will be a disgrace to yourself and all your family." **(!)** [Exclamation mark is mine, since I can't believe that his father said this dispassionately.] [Surprisingly no historians found in these ingrained behaviors any attributes that would also portend some important trait for the future!]

IV. Darwin's Health

Darwin appears to have had a chronic illness from an early age. If it were lactose intolerance or Crohn's disease both of which cause gastritis, cramps and flatulence that lasted several hours, either one may have been difficult to diagnose at that time. The symptoms fit with either illness but others as well. There were periods of remission where he could work and do as he liked. Crohn's disease is an autoimmune disease for which there is no cure, only surgery to excise part of the bowel or palliative treatment. Many with Crohn's disease work for a living at regular jobs despite their illness while Darwin was spared this inconvenience. Against this diagnosis is the fact that patients with Crohn's disease usually deteriorate with time which was not true of Darwin. At that time, lactose intolerance could have been avoided by avoiding food (especially milk and cream) with lactose in them. Unfortunately, his loving wife often fed him his favorite creamy desserts and nightly gave him warm (lactose containing) milk! There was a rumor that he was being poisoned but there is no need for such a myth when his wife was doing the same job with all those goodies, and warm milk!

One of his successful treatments was a "water cure" which did him worlds of good, perhaps because he was not eating all those creamy desserts and lactose was not in his system. Darwin's own opinion was that he had "gastric dyspepsia."

In hindsight, psychiatrists diagnosed him as psychoneurotic with anxiety spells which could bring on similar physical symptoms of gastritis, cramps with flatulence lasting several hours. Most likely, he had chronic anxiety and panic attacks. *He showed marked insecurities most of his life which he often covered up with a superior attitude, bragging and cockiness.* One article lists all the various diagnoses that were considered: Crohn's disease, psychosomatic disorder, panic disorder, Chaga's disease (weakness and exhaustion), Menier's syndrome (because of his dizziness and faintness), lactose intolerance, lupus erythematosis, arsenic poisoning, multiple allergies, hypochondriasis, agoraphobia, Cyclic Vomiting Syndrome (with flatulence lasting several hours). [I personally feel that he had a chronic anxiety disorder, marked insecurities that increased with age as the primary diagnosis with secondary lower bowel symptoms.]

However, in his day there was no science of psychiatry and many diagnoses did not exist. If psychiatry were a profession, (as today?!),[11] there would be several diagnostic categories to include. The official categories today (they change like the weather) are:

I Primary Psychiatric Diagnoses;
  Secondary diagnosis
II Personality and Character Disorders;
III Medical Diagnoses (see list above);
IV Environmental and Other Factors.
V General Average Functioning (GAF)

---

[11] Categories and diagnoses are decided by committees and whoever yells the loudest that particular year makes changes their way.

One of his primary psychiatric diagnoses would have been Chronic Anxiety Disorder with possible panic attacks (from history) with marked gastro-intestinal symptoms. A Secondary diagnosis would have been Obsessive compulsive disorder.

He suffered from a marked obsessive-compulsive disorder [which I realized, after writing the first edition of this book that was never diagnosed and had tragic effects on his life.] He kept an obsessively voluminous log of his health and symptoms as well as *complaining frequently to others.* This obsessive-compulsive tendency to record things was very noticeable in his writings. He complained constantly of his symptoms (thus seeing twenty different doctors) which is not uncommon in decompensated obsessive-compulsive individuals, (often as a *masked* depression with numerous somatic complaints). This may have been due to excessive self-imposed pressures. He reported symptoms of anxiety, fatigue and perhaps hopelessness and despair which were relieved by vacations.

When he married, he did not do it emotionally but made a list of pros and cons to *intellectually* decide as obsessively-compulsive individuals are prone to do. Their emotions are extremely inhibited and don't really know what they feel, or are able to feel for others.

Most telling of his problems was his inability to think emotionally and therefore lacked deep conviction about anything. This was very evident in his lack of conviction in the theory of natural selection itself, which he modified in later editions until the revisions sounded like Lamarck at whom he never ceased to throw bombs. [The high priests of Darwinism seem to have labeled these later editions of his book as "non-canonical" as no one refers to them.]. He was also interested about every criticism and compliment in the newspaper about his book, and loved the notoriety *from afar* (in his mansion in the little village of Downe, Kent). After his book was published, he collected albums of clippings from all sources about the book and himself. In general, he was most gracious to others but there *were* others he wasn't and did not hesitant to bomb (see below). It was reported that on his deathbed he renounced natural selection altogether, though as usual, stories differ markedly between members of his own family. His son who was there said "He said it." And his daughter who was not there said: "He didn't say it." Such are the many versions about his life and book.

It is known that he compulsively kept not only extensive records about his health but also extensive records about everything. (In this respect, he was unquestionably obsessive-compulsive).

He also had a character disorder marked by subtle antisocial tendencies (self congratulatory fibs, twisting the truth and exploiting others). He probably also had an inferiority complex (loss of his mother and a critical father) masked by a superiority complex (self-congratulatory fibs, bragging and twisting the truth) and acting very cocky at times until he found fame. Then he acted humbly and gracious as a "legend of history." Despite his unattractive physical appearance, he made up for it by his gracious and amiable personality [and wealth?] which hid the true Darwin.

One noticeable trait was his dependent personality disorder which restricted his activities to being near his wife as one will note from his letters to his wife.

His dependency and insecurities can be noted in many areas. One author, Peter Brent reviewed his early life and mentions that he and his wife, a first cousin (Emma Wedgwood) "… were linked in childhood (at) the very beginning of memory" [which would suggest they were like brother and sister.] Andrew J. Bradbury, a social psychologist draws on this biography to argue that in Darwin's letters to Emma (his wife), she was "always the mother, never the child; Darwin was always the child, never the father." He gave his wife the nickname "Mammy," [not unlike Al Jolson's famous song of the same name].[12] Darwin writes: "My dearest old Mammy

---

[12] In the days of vaudeville of the 20's and 30's , this famous comedian imitated African American dialect with ridiculous flashy clothes and straw hat and darkened skin and would sing on bended knee "Mammy, How I love you , how I love you , my dear old Mam-mie! Etc. It slurs the African American community and not sung anymore.

… Without you, when I am sick I feel most desolate. Oh Mammy, I do long to be with you and under your protection for then I feel safe." Brent states that it is difficult to see that this is a **thirty-nine year old man** writing to his wife and not a young child writing to his mother. Barloon and Noyes quote Darwin's admission to Dr. Chapman of "nervousness when Emma leaves me" which they interpret as a fear of being alone associated with panic attacks.

He was treated by no less than twenty doctors, none of whom could ascertain a diagnosis which is not uncommon when somatic symptoms are causes by psychiatric, emotional or personality factors. Physicians of his day would have been loathed to assign a psychiatric diagnosis to a man of such wealth. He may have bought a mansion in the village of Down, Kent to get away from his friends due to his embarrassing symptoms that were worse around people (nausea, cramps, flatulence; a sense of inferiority with insecurities and anxieties, which brought on such systems). These symptoms got worse with age and he stopped having friends over [fear of exposure about his twisting the truth?] He even installed a mirror on the side of the house so that he could see when someone was coming and pretend that he wasn't at home (and instructed his servants to send them on their way).

He lacked self confidence [evidence of his inferiority complex] and the older he got, the less he could tolerate criticism without having symptoms of gastric upset, cramps and flatulence. He would not mention The Book of Genesis when writing his *On the Origin of Species* for fear of clerical condemnation (they bombed him and his book anyway). For years, before receiving Wallace's manuscript he felt that a man in his social position, who was rubbing elbows with the elite clergy of his day, for him to support a belief in evolution was like "confessing to murder."

Since he was raised with all the help and support that money could buy, the only other factors that contributed to his illnesses were the fact that his mother had died when he was only eight and his father seemed to have been a disciplinarian, not a loving parent for which his sisters or a nanny could not compensate.

Darwin's children faired only slightly better and sickness pervaded the family about which Darwin felt guilty because he married his first cousin. Two died while they were quite young.

With this medical history and his history of gratuitous fibs, one can see how easily he could develop a personality disorder by inflating his ego and distort the facts in his favor. [His short statue and unattractiveness might also have contributed to his antisocial personality. His use of others can be noted in his milking his friends dry for information to clarify what he had seen on his voyage; for the *On the Origin of Species,* and using his friends in the conspiracy. However, much of this was hidden with his gracious manner unless he was cocky about something he knew or thought he knew.] This image of Darwin as an anxiety ridden, obsessive-compulsive, insecure man, needing his wife as his "Mammy," hiding from friends and with all sorts of lower bowel symptoms seems totally incongruent with the statuesque portraits that one usually associates with this little man. However, these medical, psychiatric and personalitiy disorders, should not have been ignored by historians in their rush to eulogize Darwin, instead of seeing him in a more objective light. Today we tend to see him as a man of steel, of deep conviction, a thinker and discoverer, a demigod of science, a giant, instead of a spoiled, wealthy, insecure frightened little man when his "Mammy" was not around and resorted to rather devious means to blow up his ego. [How far from reality can one get by ignoring these facts from history?]

V.   Writing Skills

One of his most important skills was his ability to write. This skill was a "must" for a gentleman of high society. Its importance can be noted by the fact that it was the only means of communication available, even for short distances. It can also be attested to by the number of "collected letters" of the period. Many wrote letters

25

even to nearby friends as a means of creating permanency to what was said in lieu of writing treaties or texts. Darwin was no exception and he seemed to have a special knack for it, as well as frequent practice. [In fact, he compulsively wrote so many letters and notes, one wonders when he had time to be physically sick]. The style of writing had to be commensurate to one's position in society to be respected. I should add that he wrote compulsively about everything and anything which clogged the machinery of his endeavors.

Darwin seemed to have taken notes about everything, compulsively, especially about information that he drained from his friends while he was writing his book on the voyage of the Beagle and *On the Origin of Species*; which includes from embryonic teeth in whales to insect wings. He recorded so many facts from his friends that his book is overladen with obscure examples in every field, particularly botany, but also in marine zoology to ancient fossils and breeding. In fact, he seems to have wringed his friends dry for as much information as he could get in return for the specimens that he had sent to them on the voyage! You name it, and he had notes about it. ***No wonder he had so many notes that he could apply to a book on evolution prior to any theory.*** It explains why he spent so much time trying to cram all this into one ponderous volume, which was still incomplete after twenty years. [Was it his own obsessive-compulsive disorder that made it difficult to draw any firm conclusions and finally abandoned it in 1850 at the age of forty-one?] However he never stopped bragging that he was getting ready to publish until he was forty eight. Was it fear of clerical condemnation or lack of a complete theory? Whatever, but we do know that he did stop working on evolution and took up classifying barnacles (an admirable, esoteric task for a gentleman of his day).

[I personally feel that he had not been interested in evolution until his mid thirties and had no convincing theory as he claimed. After his first fiasco in geology, one would think that he would be more hesitant in chancing to make a second blunder with evolution but by that time, the craving for fame was in his blood.]

Darwin had not only the ability to write well but to use the data of others as his own. He would often write things that sounded like he observed them or understood their significance, when in reality he knew nothing about them at the time. When he and the captain of the HMS Beagle found by accident large bones, he had no idea that they were prehistoric or what animal they represented (armadillo) He neither knew the animal nor the implication of their findings. However, after he returned and learned what the animals were and that they were ancient fossils (he still believed in creation and Genesis), he writes in the *On the Origin of Species* as if *he had personally* recognized that they appeared to be like the smaller armadillos that they caught and ate.

VI.   Personality

He was undoubtedly amiable and gracious and made friends easily. He had a real knack for winning friends and getting along with people. [Of course great wealth and a well known influential family had ***nothing*** to do with it.] His graciousness was commensurate with what was expected of a son of a prestigious family. In many ways, he could float around high society of London as many wealthy young men of his era did, enjoying the good life. Money, family and an amiable disposition were his calling cards. But in the shadows were his uncertainties, his insecurities, compulsiveness and health problems which caused him (luckily) to eventually withdraw from this superficial life of high society when he was still fairly young (thirty-three). Unfortunately it also kept him from completing any lengthy work (it took him three years despite his notes to complete a journal on his voyage and over twenty years ***trying to complete*** *On the Origin of Species),[if in fact he had a theory.*]

Veracity was not his strong point. He was not above bragging about himself, despite lack of factual support. He was not above alluding to "facts" that changed with time (self-aggrandizement and self defense) and throwing

28

bombs at those who seemed to be gaining an edge on him (or using others to gain information for his own ends. He clearly, but graciously, manipulated others (see conspiracy below)). Except for his medical problems, anxieties, cramps, and flatulence, *fortune shined on him.*

One gets the impression that he was humbly gracious about becoming famous when in reality he craved it. Not only in evolution but he also tried to become famous in the field of geology, **his first real scientific interest.** After having read Sir Charles Lyell's *Principles of Geology*, his only formal education in science while on the voyage of the HMS Beagle, it stimulated his interest **but only in geology and he made many geological observations on the trip, but less on natural science.** He felt that geology was the way to go in science, but after his fiasco in geology, (his first major paper in geology exploded in his face when it was completely refuted) he decided that geology *was **not** the way for **him** to go!*

Privately, Darwin was a rather bigoted man. His more extreme views were his convictions about women, Africans and primitive people.

About women he wrote in the *Descent of Man:*... "a higher eminence, in whatever he takes up, than can women—whether requiring deep thought, reason, or imagination, or merely the use of the senses and hands. If two lists were made of the most eminent men and women in poetry, painting, sculpture, music (inclusive of both composition and performance), science, and philosophy, with half-a-dozen names under each subject, the two lists would not bear comparison. ...illustrated by Mr. Galton,[13] in his work *"Hereditary Genius"* that... the average of mental power in man must be above that of women."

But if his opinion of women was derogatory, his opinion of Africans and primitive peoples was even worse. In fact, according to Richard Milton's *Shattering the myths of Darwinism*, he claims Darwin was openly a racist and developed his idea of human evolution when he later recalled his observations that Fuegians (natives of Terra del Fuego) and apes in the London zoo looked so much alike. In the *Descent of Man* he states: "At some future period not very distant as measured in centuries, the civilized races of man will almost certainly exterminate and replace the savage races throughout the world. At the same time the anthropomorphous apes...will no doubt be exterminated. The break between man and his nearest Allies will then be wider for it will intervene between man in a more civilized state, as we may hope, even than the Caucasian, and some apes as low as the baboon, instead of as now **(sic)** between the Negro or Australian and the gorilla."

---

[13] Galton was relative of Darwin.

VII. Darwin's Education *(continued)*

Perhaps because his father scolded him, he did finish basic schooling. He was then sent to Edinburgh University to study medicine as he wanted to be a physician like his father. He wasn't much better there either. Because the brave future naturalist became so squeamish on seeing an operation, he left. Protagonists of Darwin excuse this behavior because he could not accept the suffering of someone in surgery before anesthesia, (despite the fact that they were saving someone's life).

In Darwin's day a man of wealth was limited in his professional choices. He could be a physician, an officer in His Majesty's service or a clergyman. Darwin chose the clergy.

VIII. Darwin's Religious Beliefs

Darwin was a very religious person and although a Unitarian, he switched to the Church of England which was more appropriate for his position in high society. He accepted Biblical creation without question, probably stemming from his early years with his mother and his nanny; his father was considered only quasi-religious. Having failed as a physician, he decided to become a clergyman in the Church of England. He was then sent to Cambridge's Christ College to complete a B.A. which prepared him for Holy Orders in the clergy.

He had read a book by **William Paley on Natural Theology** which **he states was one of the most important books he had ever read.** Paley reasoned that if one found a watch, one would know there had to be a watch-maker, and likewise, **the eye, because of its complexity, had to have had a creator as well. Darwin never forgot this phrase for the rest of his life.** However, much later, in his forties when his daughter Annie died (age ten), he stopped going to church altogether [lack of conviction? This was not unusual, as we will see.] During the voyage of the Beagle, he would often quote the Bible to the crew if he felt they needed instruction. He still had every intention to be a clergyman while on the voyage of the HMS Beagle. Therefore issues dealing with natural history may have been interesting (particularly botany after his association with Professor Henslow), but he was more interested in adventure and geology. Until he was in his mid 30's (1844 or later) **he never expressed any thoughts that he was considering evolution seriously.** And, that was only taking it *seriously*, not formulating a theory as he later claimed for such dates as 1832, 1834 and 1838. He apparently had voluminous notes from the voyage and subsequent information from prominent scientific friends (*but not on evolution which contradicted the Bible*). [He seemed to have been more interested in keeping the peace with his clerical friends until his daughter died when he was forty-one.] This seems to conflict with Darwin's later accounts. [Each account of his religious convictions differ; some look at his notes and others at what he was actually saying and doing.] W*e have only his notes and letters* which conflict with other documents. We have nothing on which to substantiate the truth of what he says, such as his *later* memoires, (which change from earlier writings).

IX.   Darwin's Education *(continued)*

At Cambridge University's Christ College, his name preceded him. He met many important clergy, scientists and prominent figures of his day with his family background and winning ways.

However, he continued to ignore obligations to discipline himself for his future role as a rural cleric (not a naturalist). Alas, the young Darwin found other wealthy companions like himself and continued hunting, shooting rats and did not seem to have any direction or interest in his own future. He continued to be very indifferent to his studies. In short he continued to be a dilettante. *As the future will show, most of his endeavors were encouraged or helped by others,* at least until he found fame. He can thank his knowledgeable scientific friends for much of his later education and for the information and insights he gained from them which he included in his book about his adventures on the voyage of the Beagle and *On the Origin of Species. [Except for the voyage of the Beagle he did little investigation of his own* except making geological observations which must have been nearby since he stayed at home most of the time.] [How many observations could he make on his own when he was "secluded" in the village of Downe, Kent since the age of thirty-three, tied to his "Mammy's" apron strings? He acted like a landed gentry with no concern for money, education or even his future. He knew that his family was very wealthy and influential and whether or not he had an education, he would still be a part of the cream of society and find many like-minded dilettantes as himself, or so he appeared to be until his first book made him appear to be an in-depth scientist. He then absorbed science like a sponge.]

In one version he received his Bachelor of Arts degree at the age of twenty-two from Cambridge University. Another version has it that while he was at school, he was fatefully detoured from a religious life by a *phenomenal opportunity!* It was handed to him on a silver platter.

The college was not limited to only religious studies. While at Cambridge's Christ College, during his second year, he met a botany professor, John Steven Henslow who, it is claimed (*or Darwin claimed*), inspired him a great deal about natural history. Despite this inspiration he learned nothing about natural history, as we will see. The professor had taken him under his wing as a favorite. Nonetheless, Darwin had no formal education in science and admitted as much. He would become, at best, an amateur geologist and much later, a naturalist when he was in his late thirties or forties.

*The **phenomenal opportunity*** was the decision of the British government to sail from England to South America and Australia and then circumnavigate the globe to establish a series of chronometric stations in different areas, but particularly to explore and chart the coasts of South America and other areas along the route.

One version has it that the British Admiralty needed a "chief naturalist," and who got the job? Twenty–two-year-old Charles Darwin who didn't even know what the word "science" meant! [Amazing what contacts will do.] It was none other than Professor Stevens Henslow who recommended him!

There is another version. [There usually are several versions to every story when reading about Darwin.] The more reliable version has it that he was hired and given the title of "unpaid naturalists" primarily to be company at dinner for the ship's aristocratic captain Robert Fitzroy. The captain was *also* from high society and wanted a like-minded soul to keep him company.[14] The last captain had committed suicide as he had no one to

---

[14] Darwin too had the bearing and manners of a gentleman and appeared to be what the captain wanted.

talk to.[15] The Admiralty knew this and did not want a repeat performance, so he was allowed an "unpaid employee" to accompany him (Darwin's father paid his expenses). In this version, Darwin *assumed* the position of "chief naturalist" and later felt that it *was* his position [gratuitous fib or self delusion?].

His father had forbid it because it was another break in his "checkered" education in becoming a rural cleric in the Church of England, but was dissuaded by Charles' uncle Josiah Wedgwood II. His uncle had convinced his father that "the pursuit of natural history, though certainly not professional, was very suitable to a clergyman." (In short, it would not be beneath his social position).

---

[15] British captains were expected to remain aloof from the crew except to give orders and it was a lonely life.

PART TWO: THE VOYAGE

I.   On The Voyage of the HMS Beagle

Darwin set sail on December 27, 1831. Darwin was out to have a good time and hadn't thought of it as being a job when it started. What a wonderful five year vacation to see the world, away from his father's admonitions; write letters, make notes, chat with the captain and go hunting on horseback, shooting up the jungles and having a great time (which he did wherever they landed)! *Luckily Professor Henslow gave him a good-bye recommendation to take a copy of Sir Charles Lyell's Principles of Geology with him so that he could learn something about what he might observe* on his cruise because Darwin frankly knew nothing about being a naturalist or a geologist, but his social position and entertaining ways at dinner did satisfy the ship's captain greatly. They became great friends. Luckily there were others aboard (naturalists?) that could lend Darwin a hand when not busy with sailing the ship and providing support on land. They also accompanied him on his frequent hunting and shooting trips on horseback. [Naturally, they must have done all of the mundane work that Darwin didn't expect to do like cleaning, packing, crating and sending specimens to where Darwin instructed them]. In the meantime, he wrote letters to friends about his adventures and what he saw. In fact he wrote so many letters describing his adventures that friends and the public alike were excited to read about his travels in unexplored places. In short, he was a public figure even before he returned to England. When he actually returned, he was treated like a celebrity by the public. He was becoming well known.

They first arrived in Brazil and after some exploration went to Argentina.

What started out to be a pleasure cruise for Darwin, turned out to be a very interesting voyage. Darwin began to take a real interest in what the trip was all about. The *geological* wonders on the way made the trip very interesting and he loved going on day trips with the captain and seeing the wonders of the Argentinean countryside. He did learn a lot from Sir Charles Lyell's book on geology and was excited to see a volcano as described by Sir Charles.

One should not forget that Darwin was still very religious and often admonished the crew with passages from the Bible. He had no interest or knowledge of evolution and still believed in the Biblical immutability of species (creationism).

The ship stayed in Argentina a month charting the area, while Darwin made friends with the local gauchos and with whom he enjoyed horseback riding and the local life. Darwin's notes also show that he was just as interested in local customs and the goings-on in Argentina at that time as he was in the trip itself. [It was as if he were writing a travel log for a magazine.] [I can imagine such a magazine article: Come to South America! See the gauchos round up cattle and shoot the natives in Terra del Fuego! Drink Yerba Mate tea! Fiestas! Revolutions! One price covers all!]

During this time the captain was also taking him on field trips. It was on one of these trips in a small boat that they *fortuitously and accidentally* came across some large old bones partially washed out of the hillside. As presumably anyone seeing these bones, their interest would be aroused, and so it was with the captain and Darwin. Later, he claims he helped dig them out [or more likely he directed the crew] and had them sent back to his good friends, not only Professor John Steven Henslow, but also his friend the young anatomist Richard Owens who was to become a specialist in fossil bones. Paleontology was not yet a science. Darwin found these huge bones interesting and enjoyed collecting them, but had no idea of how to interpret them until long afterward (i.e., that they were old skeletal structures of ancient armadillos that resembled present ones). When he returned to England he had discussions about them with Richard Owens and John Gould to whom he also consigned ornithological specimens, particularly finches.

[There is good reason to believe that evolution was not discussed as it was not an issue to Darwin until much later. However Owens did explain to Drawn that the bones were from ancient armadillos which Darwin commented on in his Journal as if he arrived at this conclusion himself.] [On reading his book later, I wonder how Richard Owens and John Gould felt when they saw their observations coming out of Darwin's mouth as his own]. Botanical specimens were naturally sent to, none other than Professor Henslow. Reptile specimens were shipped to zoologist Thomas Bell. [What a wonderful way to endear one's self to friends, free of charge as part of one's unpaid *job*. Remember his boyhood "passion" for collecting stamps and pebbles? Now the voyage was becoming exciting, and a chance to add to his personal collection.] Darwin now *wants to believe* that collecting was the reason for his being on the voyage, and this assignment was expected of him as *"chief naturalist"[!]* As others have reported, he *assumed* this title and *acted as if he were the chief naturalist!* Who would argue with the captain's companion? [I have a feeling that the title of "captain's companion" or "unpaid naturalist" was beneath Darwin's dignity and "chief naturalist" was more dignified, but "certainly not professional"]. *He had no trouble asserting himself in this role* (cocky?). This, of course, did not interfere with the fun portion of the trip, though he did collect specimens along the way. After all, he had only to find and to help collect them; the crew did the hard work of crating and shipping them.

Later, the ship went to Chile, where he again enjoyed hunting, shooting and horseback riding as well as collecting specimens. However, in another more favorable article (of which there are many), he is described as totally dedicated to learning all that he could and applied himself diligently to the task of being a naturalist.[!]

While there he noticed strata that appeared to have been beaches at various levels on the hillside which he interpreted as "crustal uplifts," as if the land has risen in earthquakes as he had experience firsthand. *[These observations are extremely important because of his later fiasco in geology.]*

They then sailed to the Galapagos Island where much later (1845) Darwin claims his first interest in evolution began, (but that would change often). He was impressed by the variety of finches (their beaks) and sent numerous specimens to his friend, the ornithologist John Gould. The famous drawing of finches with their various beaks that often accompanies articles on Darwin and evolution was drawn by an assistant of John Gould. Darwin had no artistic ability. His single drawing in his notes on the *On the Origin of Species* was a crude outline to represent evolution.

Back in England, years before, a helper of an oyster digger, taught him how to stuff birds, which may have come in handy in the Galapagos Islands. However, it seems that he had no understanding of the various types or their relationships until much later when John Gould explained it to him and he could incorporate this information in his book.

David Guammen,[16] who writes on Darwin, calls this trip: "The most mythologized episode in the history of science." [I agree, as subsequent editions of the trip as reported by Darwin keep changing to imply his growing awareness of evolution!] If not mythical, it certainly is imaginatively modified because of Darwin himself who seems to *wander* from the truth in his impressions of what he saw, did, and *thought about* in later editions of the *Voyage*. [The first edition (1839) of the voyage was primarily an adventure story with descriptions of geological features, but in the second edition in 1845 at age thirty-six, the geological observations were imaginatively replaced by *natural scientific (biological)* observations and subtle hints of evolution! This would be a more convenient version for future developments.] Both in Argentina and Chile, he spent a great deal of time – of course – hunting, shooting and horseback riding.

At 23, leopards rarely change their spots. But contrary to expectations, like Einstein, he seemed to have

---

[16] National Geographic Magazine 2/2009.

had some sort of an epiphany, *or so he seems to imply much later*, and developed a keen interest in geological observations, *accredited to Sir Charles Lyell's book on geology.* [*Pandering?*] Later, he became great friends with Sir Charles Lyell (1797-1875); ***a truly fortuitous relationship for his future and the conspiracy,*** *as we will see!*

His interest, in the adventure was expressed in taking numerous notes about his travels and geological formations he had seen. He would write three later books when he finally finished his journal on the voyage. Naturally, they were also on geology, not biological science.

[Note: This voyage, (mostly at sea) was the only experience that Darwin had in his life with nature science, but was planning to be a priest while on this trip, not a naturalist.] A few years later he married and settled in Downe, Trent not far from London where he spent the rest of his life.

II.   The Journal of the Voyage of the HMS Beagle

On his return he wrote his well known adventure story impressively entitled: *Journal of Researches into the Geology and Natural History of the Various Countries Visited by HMS Beagle 1832-36* (published 1839). Having journals to copy, it only took Darwin ***three years*** to complete a manuscript for print *while* *consulting with those to whom he had sent specimens to learn more about what he had seen!* **This information made him look like an expert in every field.** He could then write as if he were the authority in making these observations of which he had been very naïve. He could describe well but had no knowledge of how or why things were as they were, since he still felt Creation was the answer. He only learned much later toward the end of the trip that the world was much older than reported in the Bible (in South Africa) when he met Sir John Herschel, an already famous scientist that Darwin began to admire greatly.

Biology in the 19th century, as well as all physical sciences was predominantly descriptive. Investigation and experimentations were still in the future. The sciences of physics and basic chemistry were much more advanced. Dmitri I. Mendeleyev (1834-1907) designed the most accepted atomic table used today (1869, revised 1871) that was so accurate one could predict missing elements that were found much later.

However in geology and natural science one was an authority if one knew names, descriptions and classifications (the latter was all the rage).

III. Darwin's Interest in Evolution

The information about his interests in "transmutation," as Darwin called it, and the dates are so nebulous (written by Darwin twenty years after the incidents) that everyone quotes them differently, *including Darwin himself!* He claimed later that, at the time, he revised the ***Voyages (1845)***, it was on the Galapagos Islands when he was inspired about the idea of evolution. Yet Darwin claims *twenty years later* that as soon as he returned from the voyage, he started working on his notes and **suddenly saw connections to evolution** (1838, another date!). At times, he would give three different accounts of when his interest in evolution started: *1) when he started writing his adventure story and saw the facts coming together 1838), 2) on the Galapagos Island with finches (1834) and 3) when writing On the Origin of Species, when he observed the fossils in Argentina 1832).* [Take your choice! Or, one can take a fourth date of 1844 when he first mentioned that he was taking evolution *seriously*]. He had already compiled a series of notes from his friends that could now be applied to a text on evolution as well which included a draft of a paper on a possible evolution based on what he thought were likenesses between natives in Terra del Fuego (Natives of the southern-most portion of Argentina) and apes in the London zoo. [This says a lot about his vast knowledge of zoology and his superhuman ability to notice things, when he sees little differences between natives and apes!] He apparently needed to push the date earlier to lay claim to being first with a natural selection theory, after the original presentation at the Linnaean Society on Evolution.

[Note: It was not until 1858 when he was already 49 that he finally came out of the closet with the manuscript which strongly suggest that it had been a rewording of Wallace's paper with only two weeks before it was presented which is another date to be reckoned with. Five dates in all!].

But in Argentina he says that he was only becoming aware that the age of the earth in the Book of Genesis may be flawed (more than likely it was later in South Africa when he first met Sir John Herschel who told him about the earth being much older). Later, he seemed to need earliest evidence possible to beat all disclaimers to the title of Father of Natural Selection. He had received many letters after the Linnaean Society presentation saying that "so and so" had written the same thing on natural selection earlier (which was true, as Darwin later acknowledged!). Actually, at the time, there were hints of evolution in his specimens but he hadn't a clue until years later when talking to Richard Owens and John Gould. He was still very religious and still planning to be a clergyman throughout the voyage. It wasn't until he was in his late twenties that his interest in being a clergyman slowly faded and his direction changed to science, which must have been a big disappointment to his father.

IV.　About the Trip on the HMS Beagle

Speaking of Darwin's trip on the Beagle, the Introductory notes of the Harvard edition continue: "....returned after a voyage of five years with a vast first-hand knowledge of geology *and zoology* [!], a reputation as a successful collector, *and most important of all, with the germinal ideas of his theory of evolution.*"[Emphasis is mine since zoology wasn't important in his *Journal* until 1845, eight years after the trip. He never mentioned evolution *once* in his adventure story in 1839.] It had taken **him three years to write it** until he could obtain information about what he had seen. When he started the trip he admitted not knowing what the word "science" meant. [How then, could he have interpreted what he saw, until his learned friends educated him?!]

[Note: it was not until 1845 when Darwin was 34 that he first started dropping hints about evolution in his second edition and it wasn't until 1858 that he came out about evolution *after* receiving Wallace's paper.]

V.　The Result of the Voyage

Since the world was still uncharted in many areas any story of exploration and adventure should sell well, especially by a talented writer like Darwin. With his fine style, new found knowledge and already well known from his letters, success was assured and his adventure book became a best seller. ***As scholars have noted, there is not one mention of evolution in the entire book!*** Their claim is that it was his "little secret!" [Or dare I say *"yet undreamt of idea?"*]. But *later* he managed to *reinterpret* that voyage on the Beagle to his advantage to substantiate his *"previous observations,"* when he needed more than *"yet unpublished"* data. The *Voyage of the Beagle* will have *new meaning,* [undoubtedly helped by his youthful habits, which I feel was one of the youth's greatest assets for his later fame.] The ***Voyage*** went to five printings and in later editions, Darwin made many *revisions*. One can now pick up very subtle hints that he is alluding to his insights into the association between extinct fossils and living creatures. Even fourteen years later in *On the Origins of Species,* he reinterprets his trip to have been a revelation on the struggle for life among species and its effects on evolution and ***his interests in evolution started in Argentina!***

VI.   Darwin's Life *(continued)*

At the age of thirty, he married his first cousin, the wealthy Emma Wedgwood and started raising a family. Two reasons on his list in favor of marriage were to have company in his old age and to get out of "grimy London house." At the age of thirty-three he bought a huge mansion in Downe, Kent not far from London to have room for his wife, children, servants and ample room to entertain and live as a country gentleman ["in seclusion"?]. A family proved to be no hindrance to his writing as his inherited wealth was ample to have servants and a nanny to take care of the children and his every need. His wife doted over him and he became very dependent on here and later called her "Mammy."(See his letter to his wife above under Darwin's health). Yet, according to the author of the Introductory and Historical notes, in the Harvard Classics, goes on to say: "With extraordinary courage and endurance he took up a life of seclusion[?!] and methodical regularity, and accomplished his colossal results in spite of the most severe physical handicap. .....and for the rest of his life he suffered constantly, but without complaint." [!!]

[First of all, I never knew it took "extraordinary courage and endurance" to buy a huge mansion in the country for a better life for himself, his family and servants to attend to his every need, especially when money was no object. He could, and did, live with all the comforts available and entertain friends originally (until his anxieties, pain and flatulence that lasted several hours may have driven him away from society and later "seclusion" (with wife, family and servants).] He enjoyed gardening [which now becomes "research into botany"] and he can quote many botanical references as his own that he gleaned from his many botanical friends and spend time writing when he felt up to it which was quite often, if his voluminous production is any indication. One assumes he also had a huge reference library in the house to consult when he was not in London to pry information from his well informed friends and authorities of his day.

Secondly, "but without complaint" is a more gratuitous fib by the author of these Introductory Notes than many of Darwin's gratuitous fibs. He wrote to Joseph Hooker in 1865: "I have been (for) five or six wretched days miserable from morning to night unable to do anything." One year prior, Darwin wrote: "For five months I have done nothing but be sick." (!) (There are numerous other references to his medical complaints). [Otherwise how would the author of the Introductory Notes know that he "suffered constantly," written some fifty years later?]

Perhaps I don't understand what the words "without complaint" means. [Harvard man much smarter].[17] [This author seems to be so ecstatic about Darwin that he has lost all objectivity. But he is one of many trying to outdo each other with exaggerations and gushing superlatives and telling omissions.] [Furthermore, "with methodical regularity and achieved his colossal results..." is going over the edge, since we only have Darwin's word for

---

[17] There was an old college song that went "Don't send my son to Yale," the dying mother said. "Don't send my son to Harvard, I'd rather see him dead..." which I doubt was related to reading the above Introductory Notes.

what he is doing. Actually between weeks of water cures and other health treatments, vacations and work on barnacles, it certainly does not sound like a man dedicated only to evolution. And, it may be that this pampered "always the child" little man may have felt that *any effort* was Herculean (bragging?). And, the "colossal results" did not appear until Wallace's paper arrived.]

He had every intention of moving out of London after he married. Darwin's illness was never diagnosed, but his symptoms of severe abdominal pain, vomiting, cramps and flatulence were well known from his own complaints [and friends?]. Plus, he enjoyed gardening, seeing friends originally and taking trips to London, especially for lectures and to bleed his friend pale for information. This picture of self-martyrdom hardly fits a man who in many ways seems to be enjoying gardening, his wife, ten children and servants in "seclusion" very much. Therefore, he seems to have had several good reasons for moving out of London, for love of gardening, entertaining friends originally, writing when he so desired without the presence of high society and clergy so near [...and because of flatulence that lasted for several hours at a time?].

[Note: None of the illnesses described caused constant suffering which is clearly an exaggeration. Others live and work with fairly normal regularity, so why is Darwin the exception?] About 1842 Darwin age thirty-three, continued to have **bouts of illness** with a great deal of pain, cramps, and flatulence [need I repeat for how long?] and possibly because of it, and wanting a mansion in the country, [not just for a better environment for his family and servants but for fresh air?]. It was in the little village of Downe, in Kent about eighteen kilometers outside of London, where he would spend the rest of his life gardening and writing, but could travel to London for visits as well. [How many can retire at thirty-three and buy a huge mansion in the country, servants and all the comforts of wealth and go for vacations to the seashore when he felt exhausted?]. [Who doesn't feel exhausted and need rest on the Black Sea, away from the winters in Moscow, (not to mention St Petersburg or Siberia? Answer: About two hundred million people, here in Russia!]

Later the same Introductory notes state: "His custom, which was almost a method, was to work till he was on the verge of complete collapse and then to take a holiday [to the seashore], just sufficient to restore him to working condition." [Me thinks I hear a thousand violins!] [There may have been another reason for avoiding friends which may help explain his increased anxiety on seeing others. *Could it be that he was increasing afraid that he would be asked embarrassing questions about the various inconsistencies in his dates and ideas as time went on?*

Since he lacked any conviction in the first place and was changing a lot of his ideas, it would be hard to defend earlier works, and about Wallace?] [Could he have also suffered from a form of Victorian hysteria (see Darwin's health above) that would explain his weakness and dizziness and possibly fainting? However I can hardly see this distinguished figure as portrayed in paintings, holding the back of his hand to his forehead leaning backwards and saying: "Mammy dearest, I feel faint! I should need yet another seashore vacation (and swoons onto a convenient couch.]

In the meantime, Alfred Russel Wallace lay in some obscure island hut in the jungles of Borneo, wracked with severe pain, high fevers or wrapped in blankets with severe chills from malaria (which, granted, too has periods of remissions). He still managed to work, collecting specimens for others. He had no choice but to support himself without any of the comforts of Darwin, or the money for vacations to rest, **and he still managed to write his thesis on natural selection ahead of Darwin.** Yet I hear no such superlatives about his efforts in any text![18] [Maybe he didn't "suffer constantly, but without complaint," loudly enough?] Or perhaps, he was too isolated for anyone to hear him, being often deep in the jungle with only wild animals, snakes, flies and insects

---

[18] This attitude is rapidly changing because of articles that have been critical of Darwin and hint at the important contributions of Wallace.

for company (and no wife to feed him all those goodies and warm milk every night).

VII. The aftermath of the Voyage of the Beagle *(continued)*

Darwin, having been helped by his friends Professor Henslow, John Gould, Richard Owens and Thomas Bell (in return for his help to them), he was finally able to publish his adventures on the Beagle. He could now rub shoulders with the biggest names in science as a "successful collector," and now a bestselling author of *geological research*, as an equal. The voyage made him an authority of sorts primarily in geology [thanks to Sir Charles Lyell's book on geology, but not yet on natural science.]. This was in spite of his lack of any formal scientific education. Luckily, the geological or natural sciences weren't very advanced in those days; mere descriptions sufficed (although, an important first step).

[It appears to me that Darwin realized that previous to the voyage he was a rather superficial dilettante, hunting, shooting rats and showing no interest in anything significant. Now he had recognition in his own name (not only because of his wealth, family or position in society), and I believe that it made a profound difference in him. Unfortunately fame seems to have gone to his head and he now had a craving for more recognition and fame, - at any price?] However his previous attempt at fame was to end in a fiasco.

There was a famous issue about the geological strata in Scotland in an area called Glen Roy. There were three distinct erosions in parallel along the sides of the abyss and no one could figure how they occurred. Soon after returning to England and hearing that all the big names in geology were there (including Sir Charles Lyell), he headed to Glen Roy, feeling sure he had the answer and fame would be his. He had seen similar strata near the ocean in Chile and *knew* these had to be of the same origin.[19] Feeling sure that they were the result of his theory of "crustal uplifts" to explain the formations after viewing them, he wrote a paper to that effect. To get the maximum recognition with this paper and fame for his theory and insight, he presented it at the prestigious Royal Society instead of the Geological Society which was not as important. [Did his name allowed him to present it there?] He had no scientific recognition before that time. After reading the paper he was made a Fellow of the Royal Society. [Of course he never saw glaciers in South America or any other tropical country they visited, but with his limited experience, and having read a book on geology, he seems to have felt that he knew all there was to know about layering strata.] When the Swiss

---

[19] This incident was covered rather completely by Bruno Maddox in Discovery Magazine November 2009.

geologist Louis Agassiz in 1840 showed they were due to glacier formations, Darwin's paper dropped like a lead balloon. It bombed out! He stubbornly refused to accept proofs that were published shortly afterwards. He bombed Mr. David Milne for exposing his paper as incorrect. Milne got the full blasts of Darwin's bombs, by calling him "an audacious son of a dog." I feel Mr. Maddox wanted to avoid the word that Darwin might have used instead of "dog" since the dog had to be female and is often called a "bitch."[If I may be so bold as to suggest this possible substitution.] [If there were any Nobel prizes to be had, he wouldn't have gotten one, unless he had friends on *that* committee too who could manipulate the results in his favor.] The next time he tried for fame, he was not deterred by this past fiasco, but threw care to the wind!

His new position in society helped gain even more information, not only from indebted friends, but later from the scientific community at large where he now had wide contacts. When trying to write *On the Origin of Species* it became a double edged sword. [I feel that by that time he was so overwhelmed with notes and facts that he couldn't take much more and may have been the reason he was having trouble putting such ponderous facts in a rational order, and after publishing, the facts were still in somewhat jumbled juxtapositions, [In order to sound erudite or because his obsessive-compulsive behavior made leaving facts out impossible?] and to give notoriety to many of his friends.

To readers of his day, (today too) he must have come across as an overwhelming genius of science with so many facts in so many specialties and with such detail. One must have gotten the impression he was an in-depth authority in many fields, *but it was good friends who he siphoned out and drained for the material that was in his book for which he occasionally acknowledged with warm kudos* (as well as reading all he could find and hearing lectures in London by renowned scientists). But, much was presented as his own work.

[Remember, the only real experience he had was on the voyage of the Beagle over twenty years earlier, where he apparently was primarily interested in less controversial geology, adventure, and being a clergyman and *evolution was not in the picture*. Much of the time was spent at sea where he was constantly seasick. In order to amass so many facts, it had to come from friends. *After age thirty three, he "secluded" himself in the village of Downe, Kent for the rest of his life.* [How much could he have investigated after that while tied to his Mammy's apron strings?]. [All his surmises and conjectures were obviously his voicing issues that other expounded (such as the explosive evolution of angiosperms which did not fit his theory of natural selection.] [Are we to assume that he explored this issue in his garden or library? Such research requires enumerable archeological finds of primitive plants and their relationship to each other. Are we to believe that he was doing this type of research in his backyard?] Where did he hear about it, except from others doing the research?

Despite his lack of any formal training in science, his family's name and wealth and amiable personality opened doors to famous people that gave him the information he wanted that would have been closed to others. So many people defer to wealth and love to rub elbows with the rich and famous [Professor Winslow?], even if money doesn't pass their way. (Hope springs eternal!)

After going into "seclusion," and working until he was "on the verge of complete collapse," it still took him, over *twenty years* to only partially complete this fabled manuscript *On the Origin of Species, (with so many obsessive-compulsive facts to include, rewrite, change, correct etc).* *[Could depression and exhaustion caused by his decompensated obsessive-compulsive disorder cause his doctors to suggest that he take a break at the seashore?]*

[Note: In 1850, at the age of 41, he left the evolutionary fray to concentrate on the gentleman task of classifying barnacles!] [Yes, barnacles!] [He apparently felt a despair of ever going anywhere with his

manuscript on a theory of evolution, and wanted to obtain **some sort** of reputation for having accomplished **something** during these years in "seclusion."] In his day one could become famous for collecting or even classifying things. Karl Linnaeus, a contemporary of Darwin in Sweden had done a Herculean job completing a text on classifying all known amphibians and reptiles). Such an esoteric classification as barnacles was bound to raise eyebrows among his elite friends when they wanted to know what he was doing. And, he could do it at home! He could buy specimens from people like Alfred Wallace who were working out in the field. [In the *beginning, it may have looked like an easy project but for obsessive-*compulsive Darwin it proved to be overwhelming as well.] After four years of exhaustive work he returned to write shorter papers on geology and other esoteric projects, but apparently his book on evolution laid collecting dust, despite the urging of friends to complete it and publish. [I tend to feel he may have developed a mental block against it and didn't want to get involved again because of his obsessive-compulsive needs that would overwhelm him again especially if he really did not have a theory as he claimed nor could his scientific friend suggest one as they had none themselves. I assume that his friends didn't ask questions when he implied that he was still putting his notes on evolution together and would eventually publish to allay their concerns.] Besides barnacles, Darwin was working on other projects totally unrelated to evolution, even making presentations before the Linnaean Society that is until Wallace's manuscript arrived (1858). Then, of barnacles, and geology, we hear no more!

That is, until *On the Origin of Species* seemed out to impress the reader with his minutia concerning in depth knowledge of science of which barnacles would serve that purpose.

VIII. The Art of Collecting

Eighteenth and nineteenth century Europe had not developed a tradition of laboratory research, in the natural sciences and the main concentration was on a natural first step, that of making observations. Collecting and cataloguing specimens was very much in vogue especially in the latter century in science which brought fame to the collector. Darwin was reported to be a "successful" collector. [Remember the comments that as a child he enjoyed collecting stamps and pebbles "passionately," (or should they have said "compulsively")?]

In St Petersburg, Peter the Great (1672-1725), founded in 1724, one of the first such collections in his Academy of Science (including a museum which included a collection of unusual specimens) as well as an astronomical observatory. Later, Catherine the Great (1729-1796) would collect one of the world's greatest art treasures in the Hermitage. Collecting was all the rage. It was the gentlemanly thing to do to avidly collect something, Such as art, antiques, and odd anatomical specimens which were thought to educate the gentry and the public.

Therefore it is not surprising that the voyage of the Beagle was designed to collect specimens of all varieties for the British Zoological Museum (not for friends of Darwin). (Undoubtedly, his friends had the privilege of first choice to examine and write about them, but must have sent the specimens eventually back to their rightful owners, the British Zoological Museum).

PART THREE:
EVOLUTIONARY THOUGHT IN THE
18TH AND 19TH CENTURIES

I.   Evolution in the 18th Century

As early as 1699, Edward Tyson had written about the amazing likeness of chimpanzee and human anatomy that he had dissected. This work started a wave of interest in evolution throughout British scientific circles. Even Erasmus Darwin (Darwin's grandfather) in 1770 had published that he was convinced that different species had evolved from a common simpler forebear. The church humorously ridiculed him. Charles Darwin was so religious that he apparently paid no attention to his grandfather since he still believed in Biblical Creation until he was in his early thirties. He was a devoted church-goer during this time and until he was forty one until his daughter died. [And, if he didn't believe in Biblical teachings (Genesis), should we conclude that he was a hypocrite for faithfully attending mass until he was forty-one? Then, at fort-nine, he joined other materialists like Karl Marx Lenin and Stalin to "bomb" the churches out of existence.]

II. Evolution in the 19th Century. Jean-Baptiste Lamarck (1744-1829).

In 1801 Jean-Baptiste Lamarck in France had written *the first modern theory of evolution*. In England, it was soundly blasted for assumptions that his theory seemed naïve because he believed improving a specific trait could be passed on to the next generation thereby strengthening that trait. [This may not have been so preposterous in the light of epigenetics and the present radical, theory that supports it.] *Even in the On the Origin of Species, Darwin gives a case of a change noted in just one generation that was inherited!* Later editions of *On the Origin of Species* became more and more like Lamarck's theory. The English translator said that Lamarck had written that organisms change because they "intended" to change (translator's chauvinism?) but should have been translated "sensed an internal urge" to change ("*sentiments interieur*"). Darwin loved throwing bombs at Lamarck every chance he got, despite his gracious manner toward most others. That was when Darwin appears to have wanted to be Father of Evolution and Lamarck was a thorn in his side.

Lamarck said that evolution was evolving to perfection (humans, a deity?). He also stated that species did not die, but change to other species for which he had no proof; only naïve observations. He said other things which sounded deistic which were not what scientists wanted in a theory (pure materialism). Yet the materialists have their own goddess called Mother Nature that does all the miracles former deities did, but not metaphysically, but through natural selection. This was certainly the belief of Darwin when he says in the *On the Origin of Species*: "It may metaphorically be said that natural selection in daily and hourly scrutinizing, throughout the world, the slightest variations, rejecting those that are bad, preserving and adding all that are good; silently and insensibly working, whenever and wherever opportunity offers, at the improvement of each organic being." [How, I wonder, could he say this without invoking the Goddess Mother Nature?] [It reminds me of Saint Nikolai (alias Santa Klaus) distributing gifts to good children and coal to bad children all around the world in a 24 hour time span.]

Alfred Russel Wallace (1823-1913)

Alfred Russel Wallace was a naturalist and a genuine investigator. His family had been middle class but has lost all of their money. He had to go to work after only 6 years of education. Unlike Darwin, he had no money, no influential friends and had to work at various positions such as a surveyor, a teacher and other low paying jobs for a living which took up much of his time. He was not in any "clique" where others could help him. He was a brilliant and astute investigator, (far more than Darwin) and certainly far more industrious ***considering the little***

time he had for research and writing. But for lack of background and funds he had to hire himself out to work for other investigators who had the money to pay him for collecting specimens around the world, including Darwin. He saved enough money with a friend to sail for Brazil to do research in natural history. There, he supported himself collecting many specimens in the Amazon jungles for other investigators in England and even sent specimens to Darwin with whom he occasionally corresponded. Unfortunately, a major portion of his specimens and his notes were destroyed when the ship carrying the material caught on fire and sank. Wallace was an optimistic young man and was undaunted. He continued to make a living by collecting specimens, by going to Indonesia to find new specimens for his clients and did extensive research. He had eight years of documented ***long continued observation of the struggle for existence that he may have long reflected upon*** while living in the jungles of Indonesia and the Malay archipelagos, collecting specimens for collectors back in England as well as writing numerous articles on biology. At the same time he was in touch with Darwin for whom he was furnishing specimens for his collection. They occasionally corresponded with each other and Darwin (probably through his servants) sent him a book on evolution by a man named Chambers called ***Vestiges.*** Darwin had thrown bombs at this author as well, but Wallace was so impressed that he compiled some of his own notes on evolution. Shortly thereafter, 1855, he wrote an essay*," On the Law Which Has Regulated the Introduction of New Species,"* published later that year.[20] *This article basically outlines the theory of natural selection! He stated that 1) all species come from closely related pre-existing species. 2) The new species arise by progressive divergent variations. 3) These new species survive the older species in the struggle for existence.* [21] Wallace also writes in his article (in 1855), "For ten years, the question of *how* changes of species could have been brought about was rarely out of my mind." Wallace's interest in natural selection can be ascertained to at least 1845 *by his own published account (not a secret, later revealed!)*. (Later, Darwin apparently felt that he had to beat that record). Nevertheless, in 1855 Wallace **had already published on natural selection** while working in Sarawak, Borneo, so Darwin had to fudge results by referring to "yet unpublished note going back to 1837! Presumably, because Wallace's article was not published in a prestigious journal, the scientific community ignored it. [That is, the upper class strata in science where all the important discoveries are supposed to be made in a class society.]

---

[20] Encyclopedia Britannica Vol. 19, p-530. 1959Ed.

[21] Encyclopedia Britannica, Ref: Alfred Russel Wallace, 2010 Deluxe Edition.

He had already written on evolution to Darwin who replied: "This summer will make the twentieth year since I opened my first notebook on the species **question**," [Emphasis mine] then added, "It might take two more years to go to press." [A subtle hint that he was ahead of Wallace and Wallace should give up on trying to find a solution and just collect specimens?] [Wallace may have felt that his 1855 publication was too limited! He had not made much of variations in traits in a changing environment where selection by struggles would weed out the less adapted species, [probably because the environment is pretty steady while evolution is still progressing.]

Unfortunately, years later, he wanted to expand his theory into a holistic theory that included awareness and consciousness as well. In his day where materialism had rejected awareness and consciousness as a legitimate field of research, he had nowhere to turn except to metaphysics. Here he ran into a quagmire of pseudo-sciences that caused a bizarre turn in his thinking which tended to discredit him. Nonetheless he continued to do brilliant research in natural history. Darwin, on the other hand, never really mentioned the early evolution of the mind, but only started with comparisons of humans and primates. [Dare I note that what was not in Wallace's paper was not in the Origin of Species either, since Darwin's headings were the same as in Wallace's paper?]

At that time, Wallace was not thinking holistically and there was nothing of that sort to copy. Darwin, rather than exploring new fields, later, was cashing in on his reputation by rehashing *On the Origin of Species* in *The Descent of Man* and other similar volumes [of little originality.]

III.   Vestiges

Another book was published anonymously in 1848 entitled *Vestiges of the Natural history of Creation.* It was a sensation but also condemned by the British Association for the Advancement of Science who lobbed bombs at the author for not having explained *how* evolution may have occurred. The Church of England and others condemned it with their own supply of bombs as a refutation of the Bible. During the 1840's Darwin still accepted the immutability of species. Darwin, at *age thirty-nine* in 1848 [now the "expert" naturalist], didn't hesitate to lob bombs at this author either, and disparage the tract by saying, "that strange unphilosophical, but capitally-written book. …. (the author's) geology strikes me as bad and the zoology far worse." Yet Darwin had published nothing but an adventure story and papers on geology and diverse tracts, and said that writing on evolution is like "confessing to murder."

Wallace, on the other hand, was deeply impressed by *Vestiges* (sent to him by Darwin or his servants, perhaps for sending Darwin specimens). Wallace then wrote a tract on evolution himself in 1855 (see above). Despite Darwin's notes on evolution, he apparently had no intention of publishing without a clear-cut theory because it would alienate the clergy and his friends in high society at that time "like confessing to a murder," Darwin said, because of the way that the author of *Vestiges* had been vilified. (Later, it was discovered to be a man named Robert Chambers who was blasted just as much as the anonymous author had been).

IV.   Darwin and Evolution

Work on evolution was moving rapidly and many had published similar, *if not identical, theories on natural selection* in 'out of the way' journals (to be discussed below when Darwin felt he had to acknowledge these authors and devoted most of the Introduction to the *On the Origin of Species* to acknowledging them).

Struggles

This period between the voyage on the HMS Beagle, his fiasco paper in geology and the publication of On the Origin of Species, events are extremely confusing because there are so many loose versions from apologists and others and perhaps due to Darwin's own account and history of these events and dates which seem to be in conflict.

Darwin said that in 1838 he read Malthus on populations "for amusement." Now he sings a different tune with his "long-continued observation" on the struggle of life."

Many years later (in the 1870's, some thirty years later) in *hindsight* when he was writing his memoires he writes: "being well prepared to appreciate the struggle for existence which everywhere goes on ***"from long-continued observation."*** [!] [Presumably these long continued observations were from the windows of his house in Downe, Trent?] of the habits of animals and plants, ***it at once struck me*** [!] that *under these circumstances favorable variations* would tend to be preserved, and unfavorable ones to be destroyed [bold type and emphasis and exclamation marks are mine.] "The results of this would be a new species. Here then I had at last got a theory by which to work."

It appears that he read Malthus later than stated, and it was only after corresponding with Wallace that, he became interested in struggles in evolution since he had little experience on his own. (Note: Darwin had only cursory experience with the wild (the voyage *twenty-one years* before, mostly time spent at sea when he wasn't even thinking of evolution or struggles) and certainly not in later life secluded in his mansion in the little village of Downe, while Wallace, on the other hand, *lived* in the jungles of Brazil and Borneo for more than eight years with long continued observations of struggles all around! [We know from the journal about the voyage on the Beagle, how easy it is for Darwin to voice the information of others as his very own observations.]

Where then, did Darwin get this experience of "long-continued observation," except from corresponding with Wallace or Wallace's paper? There is no mention of this struggle until after he received Wallace's paper.

[Struggle for survival or variations might have been conjectures for a theory but I'm not sure if he had really worked out the theory of natural selection as well as he claimed, or did he just lack conviction in it?]

[Actually this struggle is unworkable, because if these observations were correct, there would be no unfavorable species left as the favorable species would have eliminated all of the weaker ones and evolution would have stopped millions or billions of years ago, unless we are talking about variations within the species cannibalizing each other! How would one explain various grazing species that live together, even during famines? More appropriately there is *competition* which can lead to the vanquishing of the other species by competing favorably, but these are more the exception than the rule.]

Darwin continues in his later memoires: "...in 1836 I immediately began to prepare my journal for publication and then saw how many facts indicated the common descent of species." [!] [Didn't he go first to Glen Rock and give his "sure to be famous" paper on beaches and crustal uplifts?]

In July (1837), (so he claims) *"I opened my first note-book for facts in relation to the Origin of Species, about which I had long reflected, and never ceased for the next twenty years."* [Emphasis is my own.] [Of course, "long reflected" in secret, does not spell *"answers."* Yet, in the first edition of the voyage, evolution is never mentioned, the second edition only hints at evolution and in the *On the Origin of Species it all started in Argentina!*]

[I may be reading into the above passages, but I have a strong feeling that these Memoires over thirty years later are rather self-serving to establish credentials for having "discovered" the theory of natural selection much earlier than Wallace and others, but apparently oblivious to his own previous writing which gave different dates (which still resort to gratuitous fibs. Of course, he couldn't beat Aristotle, but he could pretend that he never heard of him or other authors already in print). One wonders when *did he actually start to put facts together*, *or did it happen only after corresponding with Wallace* or when he received his paper in 1858 *and then rushed to print*.]

By 1844, thirty-five year old Darwin told Joseph Dalton Hooker, a close friend and botanist, that he had accepted evolutionary theory and had even expanded his ideas [conjectures?] in a manuscript of 200 pages *but otherwise kept it secret from* **everyone** *(see below re: Mr. Asa Gray).*

Actually he apparently did write a tract on evolution sometime later, referring back to his impressions of natives of Terra del Fuego who he felt were very similar to apes in the London zoo! With his superhuman ability to see things that other people did not notice he *saw* there was a close similarity! [So, why missing links?] However, this study may have been written later but before the Origin of Species. The paper was never brought to anyone's attention, as usual, he kept it secret. *This conversation (or letter) to Hooker is the first proof of his interest in evolution.* [Since Joseph Hooker was his confidant and one of his co-conspirators in the July presentation of the theory, any confirmation depending on Hooker seems very dubious.]

From 1844 to 1850 the facts become very hazy. Darwin still wasn't ready or willing to go public. Considering the criticism and volleys of bombs that **Vestiges** had brought on the author, he seemed not to be too anxious to get the same treatment by publishing, [if indeed he had something to publish.]

[Yet, world fame would have been his. He lived in Downe, Kent and was somewhat removed from high society, and he need not care as much about religious or

73

social condemnation (because he had already left the church because of the death of his daughter). Therefore, I doubt that he had any conviction about a solution to the problem and plagued by insecurities, was very hesitant to make a published statement after his first big fiasco in geology.] [He certainly does not seem to have had any thought about struggles and evolution which came out later from Wallace. *We have no proof before Wallace wrote to him that Darwin, in fact, had something near complete enough to publish![22]*]

[With no certainty about a theory; just to publish would have caused bombs to explode around him for no good reason. If he had a theory, he could have published it anonymously for the record as Robert Chambers had done with *Vestiges*, knowing that many others were publishing, He obviously wanted fame, as we have seen above and will note again below. By classifying barnacles and being a famous collector, he would be of note in his own generation, if nothing else, so why would he do nothing about something so important as evolution? If others had published and caught the eye of the press and the scientific community, anything he had

---

[22][Darwin's apologists insist that he was still completing his book when Wallace's manuscript arrived, but fail to say why Darwin claims to have already sent a full description to a Mr. Asa Gray in America with the theory spelled out in 1857! Why had he waited a full years before presenting his supposed abstract. Or is this letter to Mr. Asa Gray just Darwinian mythology? Again, everything revolves around Darwin's word, not supported by objective facts.]

ever written or published *after such events* would be collecting dust today in his attic, not giving him world acclaim!]

Darwin certainly knew this and it could happen at any time! One has to conclude he didn't have what others thought he had and he was paralyzed to act. Sooner or later he would be pre-empted and he must have known this. [Therefore it is clear that in reality he had no complete theory and waited for the inevitable. He could then act the tragic figure and never have to prove anything.]

Hinting of working on a theory to close friends may simply be a throwback to his bragging and self "gratuitous fibs" which we have proof persisted long after childhood (e.g., implying that he was chief naturalist, and the dates of his interest in evolution, etc.). Darwin had published other innocuous works on geology which showed no reticence on his part to publish, *but why not on evolution?* **[Presumably, the reason is that he had nothing substantial to publish.]**

This is substantiated by the fact that in 1850, at the age of forty-one, he left working out a theory of evolution completely and devoted his full attention to classifying barnacles! However, he must have told others that he was still working on evolution and they expected him to present something any day. [This fact is conveniently omitted in most eulogies of Darwin].

[Therefore, were such "yet unpublished" (and secret) manuscripts still incomplete when he might have needed to produce such papers? More than likely, by this time he had at least hundreds of pages of notes that he was compulsively trying to fit together. By not completing his book and publishing, he was taking a huge risk of being a "might have been." Perhaps, he had no choice.]

Actually, even after he published *On the Origin of Species* few read his book except in educated scientific circles and it made little stir until the twentieth century except in intellectual and religious circles and the press (where bombs flew incessantly and the war still hasn't ended).

[It is my contention that Darwin despite his "extensive notes" and associations with the top scientists of his day that neither he nor his friends had a convincing theory of *how* evolution occurs (especially on struggles and evolution), or, his friends could have written on evolution themselves.] Darwin seemed to have ignored that evolution might be in terms of a struggle for existence or variations fit in the scheme somehow, but without a solid conviction in this theory, it wasn't going anywhere. However, he continually hinted that he was working on "his theory" [self-serving gratuitous fibs to make him appear important and on the leading edge of science?]. And because others were criticized for not saying *how* evolution occurred, he was not about to

76

publish a tract without being sure. [Darwin knew the growing importance of evolution and *if* he had a solid conviction about having found a solution, would he have been able to put it all together without endlessly writing, editing, changing, adding and taking away to complete such a lengthy manuscript? Had he been able, there is no question in my mind that he would have thrown care to the wind about clerical opinion and his friends in high society and let the bombs fall where they may and publish.] *This is strongly supported by his rush to publish **after** receiving Wallace's paper.* But when his feet were put to the fire, he had no choice but to push himself to complete his book after the publication of the Linnaean Society meeting [despite all of its inconsistencies and ponderous facts, and gratuitous fibs.]

Note: During this period evolution was all the rage and one theory after another was being debated refuted and argued. In this milieu, conjectures, convictions and theories as well as refutations of evolution were being argued. [I am convinced that Darwin with his chronic anxieties didn't know which way to turn and probably for that reason as well, he quietly withdrew from the evolutionary debate in 1850 to concentrate on his new endeavor of classifying barnacles, a less controversial and less anxiety-producing task. However, the main reason I feel is that he was overwhelmed with his compulsive personality disorder and couldn't go on, even if he had a complete theory.]

Why fiddle around with barnacles and presenting a paper at the Linnean Society on the effects of salt water on seeds at such a time?

To emphasize the uncertainty that Darwin had about evolution and any solution, we only have to remember Paley's book *Natural Theology which kept cropping up in Darwin's mind, even when writing* On the Origin of Species, where he actually **refutes** his whole theory of natural selection by referring to Paley's reference to the eye: **"To suppose the eye with all its inimitable contrivances for adjusting the focus to different focuses, for admitting different amounts of light, and for correcting of spherical and chromatic aberrations, could have been formed by natural selection, seems, I freely confess, absurd in the highest degree." [!] [Emphasis is mine.]** (Origin of Species 1859, p-162) [Does anyone else see here a total lack of conviction in natural selection?]

V.   Wallace and Darwin

While Wallace was in Indonesia he had corresponded with Charles Darwin, especially on evolution. During this period both men became aware of Thomas R. Malthus' *"Essays on the Principles of Population."*

Only later did Darwin seem to feel that this was the direction that promoted evolution (possibly after corresponding with Wallace who took the book quite seriously). Both realized that there would be a growing shortage of food with an increase in population, supposedly the stronger would survive. (Later, after corresponding with Wallace, he then seems to have taken Malthus seriously (first he said that he read it for amusement) and then made it part of the subtitle of *On the Origin of Species, "or on the Preservation of favored races in the **Struggle for Life**"*).

From his own experiences, Wallace felt strongly that *this struggle* would result in new species by defeating the weaker ones. Wallace was so impressed by Malthus' book that he made it the central issue of evolution while Darwin at that time was unimpressed. However, this attitude toward Malthus changed, possibly after corresponding with Wallace.

Also, Wallace was deeply inspired by *Vestiges* (sent to him by Darwin or his servants) and in 1855 Wallace published a manuscript on evolution. However, he failed to state *how* variations in one species could transmute into another, nor could Darwin because genetics was still unknown.

As previously stated: In one letter Wallace had written on evolution to Darwin, Darwin in his reply stated: "This summer will make the twentieth year since I opened my first notebook on the species ***question***," [Emphasis mine] then added, "It might take two more years to go to press." [A subtle hint that he was ahead of Wallace and Wallace should give up on trying to find a solution and just collect specimens?] (In one of the reference source above, a similar interpretation was made). [If one will note, Darwin says that he is "working on the species question", and (on evolution) "long reflected." [*Questions, not Answers. We all reflect on problems but it doesn't say we find answers.*] That he had a theory only comes out in the Linnean Society 1858 presentation and a year later in the On the *Origin of Species* and much later in his memoires where he predates them three times, when he is apparently sure that no one would question the dates. He implied he saw a connection to evolution when he saw the bones of an ancient armadillo in Argentina [truly a gratuitous fib]!

Remember, Darwin was only 41 years old (1850) when he stopped working on evolution and had taken up cataloguing barnacles and was preoccupied with this gentlemanly project and others in various fields until 1858, when Darwin was 49. At that time Wallace had sent him his manuscript!

Darwin may never have written anything further on evolution had it not been for Wallace giving him the answers or because it might threaten his position in high society. He would continue to collect barnacles, write on geology and the effects of salt water on seeds **until *1858, when Wallace's bombshell fell in Darwin's lap.***

PART FOUR: THE CONSPIRACY

I. Darwin's Reaction

The bombshell was the manuscript Darwin received from Wallace from Indonesia on **June 16, 1858**! It blew his mind!

Despite being racked with fever, chills and severe pains of malaria, Wallace in a flash of insight, still managed to write his twenty page manuscript in two days on *natural selection*.(Truly brilliant!). [Another reference tries to say "a brief sketch" to belittle it and minimize the conspiracy?] Wallace had entrusted Darwin to refer it to the famous Linnaean Society of which Darwin was a well known member.

[Darwin could never have dreamed that he would receive such a theory privately in the mail and it changed his world. A solution! Now events could radically change and he rushed to insure that no one was ahead of him. It was mammon from heaven!]

Since there were no copyrights in those days to establish prior authorship, publications were the only proof of origin. Wallace trusted Darwin to present his paper which would carry more weight when it was presented by a *reputable* member, -- [*so he thought!*]

Darwin, upon receiving and reading the paper, saw *the answers to natural selection staring him right in the face!* Wallace had both solutions. [Was Darwin finding a convincing theory about evolution for the first time when reading Wallace's paper?] Normal variations produced differences and struggle could weed out the lesser variations.

It had been written months before by the young Wallace but mail had to come from Indonesia half way around the world by clipper ship.

What went through Darwin's mind can only be conjectured, *but we know what he did from his letters.* Darwin had no manuscript to present. Wallace had deep convictions and his paper must have shown it. *Wallace had handed him a fully prepared twenty page document on natural selection with all the answers!*

II.   The Set-up for the Conspiracy

Alfred Russel Wallace had preempted Darwin, pure and simple, (actually by three years). What was Darwin to do? [Here, I feel, his past personal traits saved him.]

Apparently, it didn't take long for Darwin to formulate a scheme. Darwin wasted no time (*the same day!*) to rush a letter crying to his good friend Sir Charles Lyell, also a reputable member of the Linnaeus Society saying: *"I never saw a more striking coincidence"* and added *"that even Wallace's "terms now stand as heads of my chapters.*" *[!]* [The set-up] (Bold, italics and exclamation marks are mine). [***What are the chances of such a coincidence?*** It would now be possible to copy Wallace's paper as his own since he has now established how close the "two" manuscripts were as their "headings were identical." ***What an impossibly amazing coincidence!***] Darwin continues in his letter, crying: "All my originality, whatever it may amount to, will be smashed, ***but would be extremely glad now***"(emphasis is mine) (to publish an abstract of his lengthy paper.) but, ***"I would far rather burn my whole book than that (Wallace) or any man should think that I had behaved in such a paltry spirit."***(Emphasis is mine!) [Wallace's manuscript seemed to have said just what Darwin himself would have liked to have said! It may have seemed to Darwin as if Wallace had taken the very words right out of Darwin's mouth! Now they actually would come out of Darwin's mouth! Now, it appears, the conspiracy has been set in motion. He could use Wallace's paper as his own, or at least, his ideas and headings!]

[Why suddenly did Darwin become urgent about publishing? He apparently had not been concerned about writing on evolution or publishing before this. What about barnacles? What about geology, salt water effects on seeds, and other "more important" issues? Why such urgency after receiving Wallace's manuscript? (Unless, he saw answers that Wallace had expounded that Darwin

83

had not thought of by himself!). Could he have seen that struggle and variations in species, both helped them fit into new adaptations? (How easy it would have been, to slip this information into his own notes and to complete a final draft. He probably had no clear conviction which way the evolutionary question would be solved and when he saw a convincing paper, he grabbed it and wants to publish "his" theory first!]

[From analysis of his health, his anxieties and insecurities, he probably had a hard time accepting any conclusion on his own, in view of the strong controversies on all sides. He had never written anything this controversial before but when he saw a convincing paper from Wallace, he then was eager to accept it (as his own). There were no controversies on barnacles, or the effects of salt water on seeds.]

[Darwin now seems to have had his scheme in motion and the next step would be to manipulate his friends. He knew how to be gracious and humble when it was called for in high society and now was a perfect time to use his skills. After his letter to Sir Charles Lyell, he must have arranged a meeting with Sir Charles, a very famous (well deserved) member of the Linnaeus Society.]

III.   Darwin, Sir Charles Lyell and Joseph Hooker

[However, Darwin had to work fast! What if Wallace had sent copies to others in England and the word had gotten out before he could act?) In his letter to Sir Charles Lyell, he had laid the **hint** that he *"would be extremely glad now" (to publish)* but *obviously* wants to be persuaded that he "is not acting in a paltry spirit." I am certain that I am not the only one that sees a conspiracy in the making.]

[*Of course his elite friends will persuade him,* as if he needed any arm twisting? He just had to be reassured he was not acting in a paltry fashion, so that there would be no bomb-throwing later. His friends would stand by him should bombs start to explode. He couldn't afford a second fiasco (after the one in geology) to have any creditability left.]

Why had he put evolution on a "back burner" since 1850 to continue with the gentlemanly scientific project of collecting and classifying barnacles and other off the wall investigations? ("Effect of salt water on seeds." Indeed!). Is this the mind of a genius at its height?) [What had Darwin produced since the Voyage that had any sign of genius attached to it, or even read today? Wallace on the other hand continued to do brilliant research and became famous in the latter part of the 19th century.][23]

[Naturally, Sir Charles and Joseph Hooker being his social equals and good friends were going to help. They had no basis on which to deny him help since everything he did was "in secret." Who really knew the truth except Darwin himself. Why wouldn't they rely on his word, with nothing else to go by?]

IV.   The Conspiracy

[Darwin's plan was very simple as it was devious: take Wallace's paper and quickly apply Darwin's fine writing style, change it a bit here and there and - voila! – you have two almost identical papers to present, plus Darwin's would be more complete since he had a storehouse of data on his writing desk given to him by his friends, plus his fine style!]

The Harvard classics Introductory notes has one version that now, Darwin **remembers** [how convenient!] writing to a Mr. Asa Gray (one of his American botanist friends) *a year earlier outlining his full theory* [though Darwin claims he kept his manuscript *a secret*, until the full manuscript was completed, if ever]. [Did Darwin keep duplicate copies of all of his letters to others, with mimeographs and copy machines "yet unknown"?] How did he correspond with Mr. Asa Gray in the few weeks

[23] Encyclopedia Britannica, 2010 Edition. (compare with 1959 edition).

before the presentation when clipper ships were the only possibility? These details are shrouded in deep mystery. Why then, this letter to a Mr. Asa Gray in the first place, with all the details of natural selection spelled out, especially when he was preoccupied with barnacles? Wasn't Darwin a sick man in constant pain and long bouts of flatulence slaving over his new pet project of barnacles which he complained about so much? No one seems to know why he would write such a twenty page letter at this point in time, but *this **letter becomes his proof of having stated his theory in writing, and it existed before Wallace had sent him anything [so he thought, but lucky for Darwin, Wallace's paper of 1855 was practically unknown.!] (Deus ex machine, or in comes Superman (disguised as Mr. Asa Gray) to save the day!)*** Even for Aristotle, such devices were an unacceptable development in drama. This *providential* letter is another fact that seems to be swept into oblivion by the Darwinists and for good reason!

The three conspirators seem to have agreed that Darwin's paper would be presented along with Wallace's paper on the first of July, 1858.

In another more reliable text (the usual variations on a theme), it states: "The "sketches *and essays of Darwin and Wallace which filled seventeen pages of the*

Linnaean Society journal... " [24] Did it still have the same chapter headings as Wallace's paper?

In the prestigious Linnaean Society on July1, 1858, the room was hot and stuffy and no doubt members were becoming sleepy. The president of the society complained later that nothing of importance had been reported in science that year. No one seemed to know who Wallace was or cared which paper appeared first, they only knew of Darwin's name from his previous presentations and publications (the voyage, papers on geology and, of course, effects of salt water on seeds). Any discussion would be supported by Darwin's friends and the presentation would have gone completely in Darwin's favor.

The conspirators (including Darwin), were well respected and influential members of the Linnaean Society [so that no-one would suspect what was going to happen.] Then once the papers were read, they planned to use their influence on the president of the Linnaean Society to change the results which they did.

Apparently, two papers *were* presented, Darwin with the help of Sir Charles Lyell and Joseph Hooker then (behind the scenes) persuaded the president of the Linnaean Society **to publish just one paper** since they were so much alike. **They argued that because of**

---

[24] *Charles Darwin* by Doug Linder (2004).

priority interests, Darwin's name should appear first and Wallace's name as co-author! [This is despite the fact that Darwin knew that Wallace had written his paper some six months before but took that long to reach England.] Of course, the president of the society acquiesces to putting Darwin's name first, and Wallace's name a "me-too" second thus insuring that Darwin got all the credit which he did! After all, the conspirators and the president of the society were good friends and why not oblige them?

And Wallace was never informed of the decision! Wallace never knew about the conspiracy, until much later.

This version seems to be the authentic one.[25]

[They must have reasoned that no harm had been done according to what Darwin had been telling everyone about his manuscript. After all, Wallace was in Indonesia and knew nothing about the conspiracy.

Of course, now Darwin claims to have had notes and correspondence that go back to 1837 (this was 1858) *"yet unpublished."* [Would anyone else be given such

---

[25] Encyclopedia Britannica, Deluxe edition Chicago 2010. While the edition seems to be predated, its coverage of Wallace and the circumstances revolving around the publication of the Linnaean Society Journal of 1858 seems more extensive, but too brief to cover all the details. It does specifically say: "This compromise sought to avoid a conflict of priority interests and *was reached without Wallace's knowledge.*" (The latter emphasis is mine).

leeway, just on their word? I would be loathed to think that close friends and confidants of Darwin such as this Mr. Asa Gray would testify to receiving such a letter containing a full description on natural selection, *now lost!?]*

PART FIVE: ABOUT THIS CONSPIRACY

I.   Discussion and Opinion

Had Darwin's obsession to make a mark in history caused him to throw conscience to the wind? And against a man who saw him as one of the greatest scientists of his day?

It was also reasoned that Darwin had a more extensive manuscript (more embellishments and a book that would be **eventually** published), that his name should naturally go first on a combined papers.

[Since Darwin was a man of leisure, he could well afford to write more extensively prior to the meeting. He didn't have to collect specimens for others all day, so the prize goes to wealthy, leisure class not to the man who was first.]

It was said that one combined paper would be published "to resolve priority interests." **[What priority interests? Wallace was obviously first! They stole that from him!** Are we to assume that Darwin's wealth and prominent position and a member of the Linnaean society had nothing to do with it?! [Obviously, wealth had its privileges and had **everything** to do with it. In the beginning, I stated that Darwin's wealth, reputation and influence was critical in understanding how this conspiracy could have succeeded.]

Subsequently, the high priests of Darwinism blamed Sir Charles Lyell, for recommending the "deception." Years had passed and I presume that Sir Charles Lyell may have been already deceased long before the issue became more public, so he couldn't defend himself, although he did have a hand in it if only by lending his prestige to persuade the president to go along with Darwin's wishes.

Actually, **it was Darwin himself who engineered the conspiracy with his letter, lamenting his situation to Sir Charles who would not have been involved at all, if he had not received a cry for help from Darwin.** Who among them actually engineered the specific plot is anyone's guess; perhaps all three, or Darwin's own subtle manipulation of the others. (Hadn't Darwin already stated that Wallace's terms now stand as heads of his own chapters? Only Darwin knew what the truth was; the other two were working in the dark. If two similar papers had appeared, who knew the reason, but Darwin).

[Darwin had a significant hand in making the decision because he could argue that his paper was more presentable. Since they were **both presented at the same time**, no–one ever mentioning that Wallace had written his manuscript months before in Indonesia, so **[regardless of whether Darwin's paper was better or**

not, Darwin was going to come out the winner.] In any event, it was decided by Darwin, Sir Charles Lyell and Joseph Hooker who argued that **since both were presented on the same day, a combined paper would be in order**. And who would write this combined paper? Darwin, of course and put his name first and Wallace second.

And, how would Wallace know, especially after Darwin told him that he was already well ahead with his own manuscript?

[And, what happened to Wallace's original paper? Where is it now? Or has the evidence conveniently disappeared?]

[I suspect that Sir Charles and Joseph Hooker were duped by Darwin into thinking that Darwin had already half-completed a "draft" manuscript on natural selection but not yet complete enough for presentation before the Linnaean Society (reason for not having something already), let alone for publication. Sir Charles Lyell may never have seen Wallace's paper or Darwin's supposed draft, until it was read but assuming the integrity of his friend Darwin, unknowingly went along with the conspiracy. Because no one knew what Darwin was really doing out there in Downe, Trent, he could say what he wants. Sir Charles or Joseph Hooker would have no reason to doubt Darwin's veracity in view of his standing in high society and a friend.]

[Another possibility is that with Joseph Hooker's help, they convinced Sir Charles to formulate an **"innocent" conspiracy** to give Darwin what was **"rightfully his"** after the papers were presented and hide the facts at the expense of Wallace's right to fame and possible income from publishing and lectures. (Wallace could have used all the financial help he could get).]

Actually, the fact that Darwin left working on evolution eight years prior to spending his full time on barnacles says it all! *Fear of problems with the clergy and high society, fail to explain his actions because he was not apparently afraid to face social condemnation or the clergy at this point or he wouldn't have published on evolution at all!* [And bombs **did** fall!]

[Now that he was certain of the answers to natural selection that Wallace had sent to him, **he could wait no longer to publish**. No one but Wallace who was in Indonesia, Sir Charles and Hooker knew that Darwin had the paper. Darwin conspired to rush the papers so that no one could find out what was really happening *(but later despite secrecy, the facts were there to be discovered)*. If anyone found out, Darwin would be disgraced. What then, could Darwin do? **World fame would be forever denied to him.**]

[It is obvious that Darwin did not have a near-complete paper on the workings of evolution because, on receiving Wallace's paper, he would not have been so stunned (Another author has said "read in horror") or felt such a need for quick action. All of Darwin's notes looking for a possible answer would have been rubbish if Wallace were given the rightful credit.]

[If Darwin had such a paper, he could have merely announce to everyone that he had finally finished a draft of his book ready to present to the Linnaeus Society (this fantasized letter to Mr. Asa Gray?) a year earlier in 1857! Then, when Wallace had sent him a manuscript in 1858, he could have graciously had read the manuscript at the society meeting in 1858. Simple! No plot! No intrigue! However, Wallace's manuscript must have included something that Darwin found very original and upsetting which he could not shrug off and simply throw hand grenades at, as he did at Lamarck, Chambers and Mr. David Milne earlier. It appears to me that that is why he wrote to his good friend Sir Charles for help to present something **now** (despite the fact that it was already too late since 1855). His lack of any published manuscripts on evolution despite a twenty year time span to do so, tends to confirm the suspicion that despite voluminous pages to a book from conversations with friends who were authorities in their fields, that neither he nor his friends had formulated a convincing theory of evolution which to collect these facts into a coherent whole as he later claimed.]

Wallace's 1855 article on natural selection, like other articles by other investigators, were published in less prestigious journals and seldom read, so for all practical purposes they were non-extant. No Internet, no protagonists and no money or influence. But, it does exist! Fortunately for Gregory Mendel, despite being forgotten for years after his death, he was given the title of Father of Genetics.

Had Wallace been aware of the conspiracy, he could have raised a fuss, (who would listen?) but the type of person he was, he eulogized Darwin and even called his own theory: Darwinism.

Now, with fame and respect assured, Darwin could act the humble and gracious celebrity and make all sorts of kind remarks about the vanquished Wallace and write to him about "our theory," and how gracious he was in putting Wallace's name on it too! And in any future discussion add Wallace's name, now the theory was firmly attributed to Darwin. Since he was so taken back by this manuscript and its importance, **he would certainly have realized the importance of his own ideas earlier, if he really had them**. And he would have had ample time to publish instead all the rush after he received Wallace's paper. He could surely have disciplined himself enough to publish at least an abstract despite his gentlemanly fascination with barnacles, salt water and seed germination and other **important** subjects. **[But Darwin had flaws in his personality that both damned him and saved him.** He may never have continued the hard work of writing and publishing or even fully formulating his theory, had it not been for the Wallace manuscript.] To paraphrase Omar Khayyam[26]:

How many hyacinths bloom unseen,
And die, in the desert air?

[Note: In the middle of the 19th Century, England had a strong class society where the upper class looked down on and distanced themselves from the lower classes. Wallace, because he was without money, family, position or contacts, he was a **"nobody,"** so he really didn't matter. Wallace **actually worked** for a living! How much lower than Darwin can one get?]

[In elite scientific and social circles, only their social equal like Darwin would be mentioned and fame would be his. Wallace was forgotten. What the clergy thought was now unimportant though they did lobbed bombs at Darwin and his book for the next 50 years or more. He had world acclaim, despite his many mistakes and changing convictions.]

[Those that want to deify Darwin only see what they want to see. Darwinism is truly a religion and things he said and did that do not fit are considered 'non-canonical' and only wicked and callous people like Yuvenali

---

[26] The Rubaiyat of Omar Khayyam, 11th Century Persian, translation by Edward J. Fitzgerald.

Cladovainikov and myself are the true demons, the anti-Darwinites that come from Hell to bomb their god!]

Wallace was not part of any elite circle and could be ignored.

One recent article argues that Darwin is placed before the idea and has to carry the weight of this theory [after his death?] [This work on his "theory" was done at his leisure, of course, free of needing to work for a living but I assume with self imposed compulsivity, whether or not with slow infinite patience despite possible tedium.] Therefore, the author continues, we fail to see him as just a normal human flowing free without any purpose or agenda. [I wonder if a normal human being would commonly ignore the plight of other less fortunate persons toiling tirelessly (no vacations to the seashore or time out to classify barnacles or take weeks off for "water treatments" when he felt a need), while his admirer (Wallace) was still trying to make a living in the jungle with malaria and then steal his ideas, especially when all this occurred when **Darwin's own son was dying of typhoid fever!** I assume normal human being without any purpose or agenda would be more deeply concerned or in morning rather than concocting a conspiracy for his own glory. It confirms how little humanity or feelings that Darwin had with his obsessive compulsive personality disorder and self-centeredness.]

[It is claimed or Darwin claimed how deeply saddened or depressed he became at the death of his daughter Annie, but it did not seem to be associated with any somatic symptomatic such as signs of gastro-intestinal distress.]

I had never heard that Darwin might have been driven mad, had he been an ordinary human being because of the slow tedious work he was doing on evolution, **but Darwin did complain of exactly the same things while working on barnacles!** Luckily, trips to the seashore with his family (and servants?) and time out for "water treatments and other therapies," seem to have saved him.

At least Cyril Aydon concedes in his *Biography of Darwin*[27] that Darwin may not have had "the raw mental processing power of an Einstein or Newton" [and I would add Afred Russel Wallace], but still gives him A+ for superhuman ability to see things that others did not notice [like the superiority of men over women? The similarity between natives and apes? The sameness between hereditary inheritance of reduced size in one and increase size increased size of another? Gremmules? Crustal uplifts? Salt water on seeds? Missing links? The metaphor that makes the Goddess Mother Nature look like natural selection? The use of tautologies as proof? Etc.] [Why is it that so many superlatives ascribed to

[27] *Charles Darwin* (2003) by Cyril Aydon - Biography of Darwin.

Darwin fly against the wall when facts are presented?] As Thomas Huxley noted: "The great tragedy of science (is) the slaying of a beautiful hypothesis by a brutal fact."

II.   The Aftermath

When the minutes of the Linnaeus Society were published and with the well know Darwin's name first on the "combined paper," *Wallace was completely overshadowed!* Then Darwin would immediately claim "earlier letters and manuscripts, **yet unpublished,**" *dating back to 1837*, (later it was *1834, then 1832)* to support his claim of originality! [Presumably this is the reason for all the back-dating of when Darwin first had the idea of natural selection.]

In 1909, (centennial of Darwin's birth), there was much ado about Darwin in the press and of course, Cambridge University. The name: Christ College was erased and it would be renamed Darwin College. (Prophetically, "materialism" had replaced "spirituality").

The adventure story of *The Voyage of the Beagle* now becomes the voyage of *science, of research on natural selection and discovery!* His "little secret" is now revealed! Of course, no one bothers to note that indeed Wallace had actual proof of publishing on evolution in 1855 and continued to work continuously on evolution since 1845 and continued up to the time he sent his paper to Darwin in 1858 while Darwin was preoccupied with barnacles, *not evolution.* [Darwinists claim he still continued to write his book on the *Origin of Species* during this time but conveniently ignored his preoccupation with barnacles.]

Later, things were published after the facts, in *hindsight. On the Origin of Species* is the perfect example! It came out one year after the famous presentation. [If Darwin was so close to publishing, why did it take so long with all the free time in the world?] No wonder that it came along with a heavy sprinkling of kudos toward his loyal friends. Of course, no one would doubt Darwin's assertions of long years of work which I'm sure he had, but results?

You don't have to be a psychiatrist to know how people *re-edit history, and believe it,* especially when they are desperate for inheritance or world renown.

It reminds me of an incident where Arturo Toscanini, a world famous conductor, was having a conversation with an admirer. When the admirer mentioned Beethoven's 5th Symphony, Toscanini shot back: **"Beethoven's 5th?!"** **"*My* 5th !"** How easy to delude one's self when you know you would have liked to have said exactly the same thing, *if you had said it first!*

With Darwin in the limelight and hobnobbing with the scientific elite, any attack on Darwin would have been futile. How many papers criticizing Darwin and his work have been swept aside by the high priests of Darwinism? [*There were many from right after the publication until now.*] The keepers of the flame at Darwin's alter would never have allowed them to become popular by smothering them with a deluge of positive papers, journals and TV documentaries that ecstatically eulogize him with all sorts of financial supports that others cannot afford. Who would risk that for a "nobody" like Wallace?[28] Critics could attack Darwin's theories of gremmules, of crustal uplifts, slow evolutionary trees, maybe, but not the man! Any criticism must conclude that he was a genius or seek the wrath of the high priests who hold the title of "ex cathedra." The tide was entirely on Darwin's side. *Here his wealth and elite position in high society and close to all the top scientists of his day stood him well.* It happens just as often today. Professors in famous universities can make any theoretical assumptions and it stands as fact supported by his colleagues. The press goes along for the ride. [Often huge grants for the victor and his crew are in store.]

[To have taken someone else's fame didn't seem to affect Darwin in the least, though he still needed to continue to emphasize how early he had thought of

---

[28] Actually Wallace became very prominent in the latter half of the nineteenth century but not for evolution, but for other research. However, never to the degree that Darwin attained with Wallace's theory.

evolution. However, there would be no need for copyright laws today if it weren't for the many Darwins in every generation and in every field who rationalize: "I was just going to say the exact same thing, but something urgent came up (like barnacles, etc.) so I had to wait until the next time I was free. Therefore, I have every right to claim that "He was first!" *This is exactly what Darwin, Sir Charles Lyell, and Joseph Hooker had perpetrated on the scientific community and the public.]*

To show the character of Wallace, when *On the Origin of Species* was published, he even congratulated him with superlatives. Of course he was in Indonesia at the time of this conspiracy was being hatched. He only knew the actual goings on of what Darwin had perpetrated against him much later and still bore him no resentment. When he found what had happened, he kept it to himself and like the person he was, he could rise above it and even referred to his own theory as "Darwinism."

When Wallace finally returned to England, Darwin was there to greet him and gave him the royal treatment. He saw to it that he was settled with a respectable government stipend before his name faded from the history of evolution. It was the least Darwin could do for the man who *unknowingly* gave him his fame.

Despite the facts being known for years about *parts* of this conspiracy and *documented*, the Darwinian high priests defend him without caring what the facts are. This was particularly true of the patriarch of Darwinism, Steven J. Gould, who chose to see facts differently and he was brilliant enough to do so (e.g., accusing Baron Georges Cuvier of GIGO (garbage in–garbage out) to defend Darwin) when essentially Cuvier was right except for his conclusion, but his argument was an Achille's heal in natural selection.

In a rather convoluted, round-about logic to defend Darwin against Baron Georges Cuvier (1769-1832) using anecdotal documentation to prove Cuvier was using GIGO when Cuvier was more right than wrong in saying that a new evolutionary change, changes the *entire* organism which disproves natural selection and evolution. Gould felt *this conclusion refutes Curvier perfect argument against evolution.* [I agree that it *was* an unfortunate *conclusion,* while Darwin and Gould felt that single changes could be made independently in evolution.]

However, using the same type of anecdotal data, one can prove that Cuvier was right in principle though his conclusion was wrong. The evolution of a tiny nerve in the thumb muscle (the thenar nerve) allowed the thumb to close with the other finger allowing for the first time in hominids the ability to grasp things firmly, which changed the entire way of life of hominids, their tool-making skills, their diet, their biochemistry, their anatomy and nervous system, in short, everything! Cuvier was right. Darwin and Gould were wrong. There are infinite numbers of anecdotes that support this conclusion.[29]

Similarly, professors who are certain that Shakespeare wrote the plays and sonnets attributed to him, refuse to entertain any doubt, despite strong circumstantial evidence that show that someone else *could have* or *must have* written them. Books have been written about this possibility but the professors stand their ground with the same facile excuses. Undoubted Darwin will be defended the same way. First ignore, until the noise gets too loud; then defend with sweeping statements that ignore the facts. And like Professor Steven J. Gould, they are brilliant at it, except that he could also do it by finding unbelievable anecdotal "proof."

III.   *On the Origin of Species*

Now Darwin's manuscript had to be realized, despite its flaws and contradictions because now everyone was waiting for it; no more detour into barnacles or other esoteric subjects. Many have lobbed bombs at Darwin because of these flaws and contradictions but the flaws

---

[29] Some macaques and baboons have this ability but were too primitive mentally to fully use it. They were not in the same path of evolution as hominids (or, could the path be wrong?). Other primates are notorious for poor carrying and throwing skills.

have been white-washed over or the high priests of Darwinism have bombed the critics into silence, especially Steven J. Gould (who was not the first to use intellectual bombs), but Professor Gould *was* brilliant! He invented the first "silent bomb." It went off and no one knew it, one might say, (like the one he intellectually blasted at Baron Georges Cuvier.).

One year after the presentation, Darwin's masterpiece was published. Nonetheless, along with the ridiculous tautology: "Survival of the fittest"[30] (from Herbert Spencer), he seemed to feel now that this was the direction that promoted evolution (or possibly after reading Wallace's paper).

IV.   Scientific Reaction

During the year before the long awaited book *On the Origin of Species* was published, numerous letters poured in to Charles Darwin after the publication of the minutes of the Linnaeus Society meeting. He had been reading every letter and article about the presentation. After the book was published, he collected every single thing he found about him and the book which created volumes of clippings, cartoons and criticisms. [The press loved the debate and Darwin humbly and graciously loved the

---

[30] Might I ask, dear reader, is not everyone who survives the "fittest" by definition? Therefore, Natural Selection *selects* only the fittest. And, who are the fittest? Those *selected* by the" Goddess Natural Section" or Mother Nature! – (metaphorically speaking of course).

fame.] In 2002, Richard Milner wrote an article entitled *Putting Darwin in His Place* states that Darwin "clipped, catalogued and indexed hundreds of offprints, about 350 reviews and 1600 articles, as well as satires, parodies and Punch caricatures, with which he filled hefty scrapbooks..." [but Darwin was a man that "bore the weight of his theory"! He seemed to love it.]

Prior to publication, Darwin sent twenty five prepublications copies to his friends and those he respected. Many who were favorably mentioned sent back favorable comments.

However, when Sir Charles Lyell wrote (in *Antiquity of Man)* a lukewarm review of Darwin's book, Darwin had bouts of stomach pains, vomiting cramps [and flatulence?] *continuously* for ten days afterwards (not several hours!). A review of *The Descent of Man by George Mivart was quite negative. It "triggered" two months of "giddiness" and inability to work (*quoted from the same article by Richard Milner).

He also sent a copy to Sir John Herschel, his idol for years and who, he stated, inspired him to become a scientist and whose opinion he valued "more than almost any other human being." ***Sir John sent back a "bomb" that exploded in Darwin's face!*** He hardly expected it. It said that Darwin had no theory at all! Unless he could say what caused variations. It was a "law of higgledy-piggledy." If Sir Charles Lyell's review caused pain, cramps and flatulence for six solid days, we are fortunately spared finding out how many days Sir John Herschel's blast caused him.

Actually Sir John Herschel was involved in evolution and had published a rather prescient paper critical of Darwin's view stating that "useful traits could never arise from simple random variations. Evolution would always require mind, plan and design,[31] to the plain and obvious exclusion of the haphazard view of the subject and causal course of atoms." Sir John may have been thinking religiously but this view has recently received partial support from quantum physics, especially "mind."

Other writers reminded Darwin that *many investigators had* **already** *said essentially the same thing* (mentioned earlier). In order to disclaim such pretenders to the throne of "Originator of Natural Selection Theory," Darwin spent the Introduction of his book acknowledging with faint praise or refuting each and every researcher so as not to have any pretenders for his mantle of Father of Natural Selection or that he was acting in a paltry fashion.

V.   Aristotle

This included a reminder from Mr. Clair Grece who had just made a new translation of Aristotle, stating that *Aristotle had actually formulated natural selection*

---

[31] Actually, plan and design *are* mind.

2,500 years prior! Darwin claims to have been unaware of this fact.

[To say that he was not familiar with this work by Aristotle is *extremely strange* when one remembers how frequently he complained how the classics had been rammed down his throat at school. Aristotle was bedrock knowledge for any educated gentleman in England at that time. Aristotle's views on evolution, *one would have thought* would have left some impression on him. Not so! Shooting, dogs and rat catching were more important than studying.] Nevertheless, Darwin showed no interest in Aristotle or evolution until he was in his mid thirties, and no interest in Aristotle until Mr. Grece brought him to Darwin's attention. However, failure to know that Aristotle had written about evolution seems *highly suspicious,* since there were ample translations of Aristotle in his day, *besides* the one made by Mr. Clair Grece, who Darwin "claims" pointed out this "oversight passage on evolution" to him:

Quoting Aristotle (speaking of teeth for cutting and grinding,) [which I must admit is a poor example]: "....since not made for the sake of this, but it was the result of an accident. And in like manner as to the other parts to which there appears to exist an adaptation to an end. Wheresoever, therefore, all things together (that is all parts of one whole) happen like as if they were privately constituted by an internal spontaneity and whatsoever things were not thus constituted, perished and still perish."

Darwin goes on to admit: "We here see the principle of natural selection *shadowed forth* [emphasis mine] but" ... (He then goes on to "explain why Aristotle missed the mark that Darwin had so clearly revealed.)

Aristotle is not easy reading (as evidence by the above text) and often his writing needs further elaboration, but here, it is absolutely clear, as it was to Darwin as well, that *Aristotle was saying exactly what natural selection is all about, not merely "shadowing forth"*. Namely, "….since not made for the sake of this (teeth for chewing), (these parts had evolved for some other purposes but (a variation) fit into this new purpose (new function of chewing and had persisted) but it was the result of an accident" (by the chance variation of natural selection) and in like manner as to the other parts to which there appears to exist an adaptation to an end. (Similar applications of variations to new uses). Wheresoever, therefore, all things together (that is, all parts of one whole) happen like as if they were privately constituted by an internal spontaneity (today we would say gene variations). (Darwin's fantasized gremmules), "and whatsoever things were not thus constituted, perished and still perish." (Whenever variations did not find a use, they disappeared, or atrophied).

This is pure natural selection as Darwin realized! While Darwin was able to give it a greater "scientific spin," he adds little more. *Darwin had every right to call Aristotle the "Father of Natural Selection" and the "Father of Evolution" but he seems to want desperately those titles for himself.* Subsequent Darwinists (defending Darwin) state that Greek and Roman philosophers only had partial statements on evolution and not full theories. Yet Darwin himself said that ***his only real contribution was that variations in traits fit into new environments better that previous ones and persisted, essentially the same view expounded by Aristotle!*** *(or, was he just acting graciously humble?)*

Aristotle did not know that "privately constituted parts" were genetic in origin, and neither did Darwin who conjectured on a theory he called Pangenesis, where some imaginary little ***gremmules*** coming from various organs ended up floating around in the blood somewhere and eventually ended up in the testicles, were responsible! His high priests, ever after, have been desperately trying to bury this ridiculous idea. [Note: When left on his own, this "giant" in science, this unparalleled "genius" came up with some pretty off-the-wall ideas, let alone the effects of saltwater on seeds.] [One wonders what unparalleled insight got him on that tract of salt water and seeds so that he ignored evolution in order to prepare a paper which he read before the

Linnean Society while Wallace was working on a paper to publish on evolution during that time.]

We know that only after the work of Gregory Mendel (1822-1884) on the pea plants," that the concept of single units (genes) controlling variations would evolve. He is rightfully considered the Father of Genetics. While Professor Gould felt that Darwin might be given this title of Father of Genetics as well, because he used the word "genesis" from which "gene" is derived. Other scientists seem to have politely ignored his suggestion.

VI.   The Father of Evolution

The result is history. Darwin not only became the Father of Natural Selection but Father of Evolution as well! The clique of high priests who worship at the shrine of Darwinism are active, even more so today in writing eulogies, erecting monuments, naming colleges after him and foaming at the mouth trying to outdo each other with superlatives. They splash his name everywhere possible. They hide the fact that the *true modern Father of Evolution was Jean–Baptiste Lamarck*, and the ancient title should have gone to *Aristotle*.

Lamarck (Jean-Baptiste de Monet de) was *the first modern scientist* to propose a full theory of evolution and deserves the mantel of "Father of Modern Evolution" but could not identify what caused acquired characteristics to be transferred to a descendent. (Neither could Darwin determine what caused variations in traits that created a bell curve in their distribution. The British Association for the Advancement of Science should have agreed with Sir John Hershel in lobbing bombs at Darwin as they did at others for the same reason). Lamarck assumed that strengthening a given part would eventually cause this trait to be passed to subsequent generations but could not specify the internal process, since there was no adequate theory of genetics at that time where such variations could take place.

Today, experiments prove that this can happen and the issue has been for all intents and purposes been resolved. Also, *societies evolve by Lamarckian evolution of acquired characteristics, not by natural selection!*

Darwin *himself* used an example of an *evolutionary change that occurred in one lifetime that was inherited* which bombed his own theory to pieces.

In his chapter on "Atrophied organs", he writes: "But we learn from the study of our domestic productions that the disuse of parts *leads to their reduced size and that result is hereditary.*"(!) [Emphasis is mine]. *This is pure Lamarck!* If organs can be reduced by disuse and inherited by the next generation, then the overuse could lead to enlargement as well! The strength of athletes *could* result in it being inherited, if the following generations were inclined to follow the same life style. Darwin, in his "unbelievably insightful nature," saw things that other people did not notice. Yet he saw no connection to the reverse. [He only makes this connection in the "non-canonical" fifth edition of the *On the Origin of Species*, which no one is supposed to read.]

Lamarck stands exonerated on this issue though he made other claims that are not as easy to dispel. It is known that both environmental and internal influences within the organism can influence the genetic code while Lamarck assumed only environmental pressures would cause it. If he *had* said that organism change by "intention", he may be proven right after all. [We change because of our intents to change all the time, and aren't we part of the same evolution?]

A subsequent "test" of his theory was carried out by cutting off the tails of mice for 50 generations to see if their tales would become shorter as the trial went on. I presume the "scientist" was not another Ivan the Terrible, Dr Frankenstein or totally insane, but he probably could have benefited with *some* therapy. Had he been looking for *increased healing rates* as the experiment went on, it would have made far more sense, since there does not appear to be any advantage to the mice to have their tails chopped off in such an arbitrary and brainless fashion, especially if they still needed tails for whatever reason.

In 2002 the late Steven Jay Gould, the patriarch of Darwinism, had written one of the most extensive (and I might add convoluted) text to prove Darwin is not only a demigod in evolutionary theory but practically the Father of Genetics as well, even though his crack-pot theories of gemmules, missing links and time frames are some of his most ludicrous speculations, not to mention his metaphorical allusions (i.e., the intervention of the Goddess Mother Nature in natural selection).

Even his speculations on sexual evolution are debatable. He is only speculating with limited observations made by friends. It is far more complicated than any simplistic ideas suggest. Even today there are new discoveries that put new twists on sexual evolution. If Darwin had a consistent theory, why wouldn't it apply to humans? Why do some people consistently pick the same type of losers to marry?[32]

Having the deepest respect for Steven J. Gould, (my idol for years) to have written such a voluminous and contradictory work, was unbelievable to me. He even eulogizes Darwin's prose which gave Gould such comfort, as if they compete with Pushkin or Shakespeare.

Even in his best efforts to defend natural selection and Darwin himself, he seems to be repeatedly saying that while *others were more right than Darwin,* (that

---

[32] The question is rhetoric at present. Modern psychiatry has answered those questions. One seeks the familiar not the rational. Darwin could never have come to such conclusion since his own choice of mate was done without emotions but by making a table of pros and cons and deciding intellectually as many obsessive –compulsive persons do.

Darwin may not have *always* been right), but **Darwin was *never* wrong!**

VII. History

And so, Alfred Russel Wallace is buried in the back pages of history even though he was *first*. Recently, he is gaining some status as facts become known long after his death but he could not have enjoyed the fame that Darwin had in his lifetime. And what if Darwin is accepted as a plagiarist and perpetrating a conspiracy of the most disgusting kind? Will it rattle his bones? Will it have changed his life? Not one iota! What it will change is the worshiping habits of the scientific community when they realize (if ever) they bow to a fallen idol that they have to replace.

[Martin Luther while sitting on the toilet had great inspirations. Darwin too, may have had the same, (he certainly was on the toilet a good bit of the time) *but for prodding of* Wallace's *manuscript*, those ideas may still have remained constipated or ended up going *down* the toilet.]

PART SIX: MATERIALISM AND OTHER ISSUES

I.   Material World of the 19th and 20th Centuries[33]

This brings up another issue that is often attributed to Darwin; making him this giant in history. In creating this giant, bigger than life, it gives one a sense of a place to worship, as Professor Steven Jay Gould and many others seem to have found. Making Darwin a demigod may serve its purposes but is unjustified and unwarranted. That is, to place Darwin in the center or beginning of a revolutionary change in thought in all fields of science, literature and philosophy as if he or his book could have created such a momentous change or propelled it forward. Often Darwinists make this assumption with gushing superlatives to bolster his reputation about his effect on all branches of science, literature and philosophy. *It couldn't be less true. His ponderous book was hardly read until 1940 when genetics became*

---

[33] With the permission of Yuriy Alexandrovich Kalinikov, I have taken section from his book A Radical Theory of Evolution and included it here. I felt that as the one responsible for the English versions of both that Yuvenaliy Vladimirovich Kladovainikov's book should end on a positive note that this new theory represents to which Uriy Kalinikov fully agreed since the two are good friends.

a science and it fit with their misguided random mutations theory, (or rather speculations).

The material sciences have been evolving ever since Galileo first started doing objective experiments. The erroneous convictions, particularly of Rene Descartes and Emanuel Kant that anything not material was metaphysical or spiritual came to be the rallying cry of the material sciences. These men defined the material science, "materialism," as only those things you could see and touch or test by instruments. From then on, only material things could be investigated to be *objective.* This dichotomy between mind and body has persisted ever since. The material sciences have defined their turf opposed to and against the religious metaphysical dogma. The battle lines were drawn and the material sciences were winning. But, the "mind" was off limits. It was relegated to philosophy and to the spiritual. By the late 1700's most all sciences had been expunged of all metaphysical ideas, except one – biology. The Book of Genesis was still the law of the land. Humans and all living things came from a deity. It was the one area where science feared to tread. They could study plants and animals with some impunity but humans were sacrosanct. By the year 1699, a crack was made in this wall by Edward Tyson who published his anatomical dissection of both a primate and a human which had remarkable similarities. The crack widened when the ancient bones of prehistoric animals and humans were found in Europe suggesting that many species had come and gone but similar to the present ones. By 1770 there was Erasmus Darwin (Charles Darwin's grandfather) publishing that he believed that all species evolved from some primordial form. The wall was coming down and the force that was doing it was this new *idea* of evolution. It was *inevitable* that someone would come up with a materialistic theory to explain it. *It had to be non-metaphysical to be acceptable **and by some reputable person who had the support of the scientific community and the press.*** Therefore, Lamarck was rejected because he implied life forms were evolving to "perfection," (toward a deity?), as well as other aspects that sounded too metaphysical. Many investigators proffered such theories but none caught the public eye, neither the support of the scientific elite nor the press. Then *several* researchers came along proposing a purely materialistic theory of a natural selection process that could be the basis of evolution. One of them was by Wallace (which Darwin promoted, under his own name as main author). It was then published in the prestigious Journal of the Linnaean Society. However, Darwin was the only prominent (wealthy) man with contacts in all top branches of science especially such prominent men as Thomas Huxley who would champion him as did the press by giving full coverage to the issue and ensuing controversy.

Therefore to imply that this entire revolution started with one man and his book sounds more like one of Darwin's self congratulatory fibs (dare I say "fabrications"?). Except for using the book as an example for lack of any other book that received such notoriety, it did not produce the changes in society that is claimed.

But the biggest promoter of Darwin was Darwin himself and his memoirs which the scientific community drank up like borsch. The Origin of Species is dull reading and ponderous, but despite his "gratuitous fibs," his memoirs could sing!

Why was Wallace's theory (promoted as Darwinism) so special compared to the rest? The fact is, it was not, but *Darwin's prominent position in society, wealth, support from the entire scientific community and the press, were.* The theory was endorsed because of the book he had written filled with encyclopedic references that was as impressive as it was self serving, with ample kudos to his friends. *It caught on because – its time had arrived.* It was what the science community was looking for, a materialistic view of life that helped to place biology in the material sciences. The scientific community had the last piece of evidence they were looking for to vanquish the powers of the spiritual authorities and to ascend to the throne of ultimate authority themselves. It was not the man nor his book but the fact that life had been explained on a materialistic basis *that was symbolized by the man and his book. He changed nothing in reality that was not already in the wind.* And except for educated circles which now reinterpreted the place of humans in this new context, it had little effect. The church and many others continue to lob bombs at Darwin and his book but otherwise the book was essentially ignored until a new material science was born – genetics. Genetic scientists found in this reference a theory that blended with their own genetic quasi-theory: random mutation. The material scientists wallowed in their success. Their victory was complete. *Their time had come!*

II.   Summary

What was expected to be a pleasure cruise for a wealthy playboy, turned out to be a bonanza of eye opening experiences from large bones to an unbelievable variety of finches and mocking birds and all sorts of geological formations. Now, the clergyman to be, turned amateur scientist found notoriety in writing an adventure story laced with insights that he siphoned from knowledgeable friends that made him look like a powerhouse in science. It was the first time in his life he had done anything substantial and it seems to have gone to his head. As evolution raged around him, he too wanted to jump into the waters but after the vilification that others received, he was afraid to enter the fray because it was "like confessing to murder." While he hoped to solve the answer to evolution, he, nor his friends had a final convincing theory, but he gave hints that he was "working on it" and had other believing him [if he could get over his personal handicaps] that he was coming to a final version, despite conflicting statements and convictions.

And if time were the only factor, he could easily have completed his book, long before Wallace, if indeed he had a complete theory to write about. The failure to publish, or even present an abstract at this most critical time and even stopping work on evolution to classify barnacles almost certainly proves he had nothing to publish but notes that proved nothing. His friends could not help him as they too had no theory. [However, there is the questionable issue of his tragic personality disorder: his obsessive-compulsiveness which may have dragged his work through the endless quagmire of detail, rewrites, inclusions and questioning his results. Yet this disorder cannot justify completely his dropping evolution and changing to barnacles.]

[Since, emotionally, he seemed too inhibited to feel his own convictions and no insight into a theory. He could only arrive at things intellectually instead with balance.]

Wallace had both emotions and intellect and he was able to move ahead with conviction and enthusiasm, while Darwin was only egged on by hope of fame. [He could easily, in this case, have grown weary of his own obsessions and when near collapse, would be wealthy enough to take his family to the seashore for a rest. But that was not enough to clear his mind and he appears to have stopped work on evolution or go into a deep depression.] In 1850 he decided to withdraw from the evolutionary fray and became preoccupied with classifying barnacles! He may have felt it would be easy but apparently his obsessive-compulsiveness dogged him there too. [In 1858 when Wallace's bombshell arrived, Darwin must have seen it as unbelievable luck. He instinctively reverted to his old tricks of twisting the truth and fabricating a solution to his advantage, for which he had plenty of experience. Darwin quickly concocts a plan to save the day but at the expense of Wallace, for whom he seems to have had no feelings, or apparently anyone else for that matter, with the possible exception of his "Mammy," ("always the child, never the father").] It worked! The rest is history.

III.   Afterthought

If one reviews Darwin's accomplishments outside of his book *On the Origin of Species,* and one discounts his wealth and fame, and information from others, one notes that there is little that he did that has made a dent in the 20th century or beyond. In retrospect, none of his other works are of any great importance as he humbly and graciously admitted. And indeed, if what he is famous for was the result of a conspiracy, ***then***, there is little else that makes him stand out ahead of others.

No one has the fortitude to place Alfred Russel Wallace in his rightful place, because they have already enshrined Darwin as a demigod and attributed all sorts of heroic deeds to his name, *regardless of who the man really was*. They need a symbol to worship. Darwin has become that symbol.

IV.   Theories of Evolution

But let us take a closer look. Almost every day, someone is finding a new and ingenious insight and application of natural selection **which gives more prestige to their own work as well as giving the theory of natural selection and Darwin an even higher status**. It is the *genius of his present disciples* who attribute to him all sorts of phenomenon, regardless of how abstruse their reasoning is.

[It reminds me of an incident that happened in my teen-age years. I was brought up as an atheist and was now living with a very religious woman. She came home one day from church after hearing a priest give a talk about Jesus of Nazareth. She was so enthusiastic about this talk she had to tell me about it. She said that when Jesus appeared before a few of his disciples after his crucifixion; they did not know if he were a man or a deity. It was wonderful how he understood their feelings, how in such simple words he reassured them not to be afraid, not to run away but to appreciate his humanity. I wondered what these wonderful words were that were so remarkable and asked, what did he say? She replied with exuberance, He said: "Give me meat." I began to wonder if other such statements by Darwin were equally so interpreted. I wonder if Darwin's disciples are over-interpreting everything that happens under the banner of natural selection. The genius seems to be coming more from them and their ingenious interpretations than from Darwin.

The random theory of evolution is just as leaky and despite articles showing its weakness, they are overshadowed with prominent articles that eulogize it.

[However, times are changing. A new theory has been presented that is as radical as it is revolutionary and will give the scientific community plenty of bomb-throwing opportunities (see below).]

V. The Material Universe

The material sciences have set us adrift in a meaningless, purposeless universe. Materialism leads to hedonism with no moral rudder to steer our future but to be preoccupied with materialistic technology to keep us from thinking of our end. (No wonder Creationists are coming back in vogue to refute material evolutionary theories.)

In reducing us to "another animal," they have at least made us humble and appreciate the lives of other species as our kin. But now we are a civilization of irreconcilable differences.

The mixture of religious people with materialists has made us a race of paradoxes in chaos, one believing in a destiny which the other tries to refute. The misguided religionists try helping others who then overwhelm the helpers. Despite the well-meaning of this help, it disrupts more than it helps and after an initial up-serge in the quality of life, it often plunges them into greater poverty, starvation, crime and despair and materialists are inventing the wherewithal to increase this effort. Our moral institutions are failing while we are destroying stable native ("primitive"), though moral, cultures, that had sustained them for hundreds of thousands of years. And, modern faiths that *are* viable, attack one another with murderous rage; but without faiths, we drift into materialistic hedonism and despair.

VI. The 20th and 21st Centuries

But now, Aristotle, Wallace, Darwin and rank materialism are passing.

New Sciences are developing on top of the older material sciences that do not ignore our most valuable possession, our awareness, our intellect and our emotions that Wallace tried to include but failed. So far it is not a part of the material sciences, thanks to such myopic and presumptuous philosophers as Rene Descartes and Emanuel Kant. Psychologists test and describe, but fail to integrate our mind and our bodies with the universe, nor try to understand awareness, mind or emotions but merely accept them.

The new sciences are already on the way and need only a new theory of evolution that is inclusive of all aspects of life, awareness, consciousness and feelings, not just our anatomy. It has profound implications that should dwarf Wallace and Darwin and other former materialistic evolutionary theories such as random mutation. Our minds and emotions are only now being considered seriously by quantum physicists. *Quantum physicists did not set out to discover the mind, they found that the results of their experiments are such that a concept of mind was inescapable and that it permeates the universe.* A new theory of evolution is needed that takes note of this science.

My good friend Yuriy Alexandrovich (Kalinikov) who wrote "*A Radical Theory of Evolution,*"[34] which has been recently published, is a rational and consistent theory of evolution where he presents a holistic view of a

---

[34] *A Radical Theory of Evolution* by Yuriy A.Kalinikov (2009; ISBN: 1449519415), available from major online bookstores.

living evolution. It should have profound implications that should dwarf natural selection and other former materialistic evolutionary theories as random mutation and put to rest creationism which thrives on the failures of materialistic evolution.

Epilogue

When scientists had dethroned the concept of a deity, they had no alternative but to make a man the idol (or Mother Nature, which in a sense they have, but not some metaphysical being!), like socialism made Lenin (but they had to blast Stalin and other saints out of the pantheon). It was Nietzsche who was to lament: "God is dead," and now we know who bombed the deity out of existence. But like the Bolsheviks who literally used bombs to destroy churches in order to dethrone the deity, and replaced the deity with materialism, the result was disastrous. Materialism has done the same thing and fragmented the universe and people into units that one can see and test, not as religions have done for ages by uniting people and the universe and give them significance, and blind hope. Bombing and debunking Darwin and other materialists or a new evolutionary theory can never re-establish a personal deity again, which may not be necessary. Millions of Hinayana Buddhists live happy moral productive lives with no such belief. Their deep faith is in the *teachings* of the Buddha who died over 2500 years ago. He taught compassion and understanding that has been a guide for over 2500 years to millions of people who still maintain their faith today while still accepting the full concepts of evolution and science, *but* not just the material side, but a holistic universe, embodying universal awareness, consciousness and emotions in all life – everything as One.

And, we too can re-establish our moral basis in the new science as well as our own total being and significance by uniting again with something truly great: the conscious universe.

Yuvenaliy V. Cladovaynikoff